Inhaled Pharmaceutical Product Development Perspectives

Emerging Issues in Analytical Chemistry

Series Editor
Brian F. Thomas

Co-published by Elsevier and RTI Press, the *Emerging Issues in Analytical Chemistry* series highlights contemporary challenges in health, environmental, and forensic sciences being addressed by novel analytical chemistry approaches, methods, or instrumentation. Each volume is available as an e-book, on Elsevier's ScienceDirect, and via print. Series editor Dr. Brian F. Thomas continuously identifies volume authors and topics; areas of current interest include identification of tobacco product content prompted by regulations of the Family Tobacco Control Act, constituents and use characteristics of e-cigarettes and vaporizers, analysis of the synthetic cannabinoids and cathinones proliferating on the illicit market, medication compliance and prescription pain killer use and diversion, and environmental exposure to chemicals such as phthalates, endocrine disrupters, and flame retardants. Novel analytical methods and approaches are also highlighted, such as ultraperformance convergence chromatography, ion mobility, in silico chemoinformatics, and metallomics. By highlighting analytical innovations and new information, this series furthers our understanding of chemicals, exposures, and societal consequences.

Inhaled Pharmaceutical Product Development Perspectives

Challenges and Opportunities

Anthony J. Hickey
RTI International, Research Triangle Park, NC, United States

Elsevier
Radarweg 29, PO Box 211, 1000 AE Amsterdam, Netherlands
The Boulevard, Langford Lane, Kidlington, Oxford OX5 1GB, United Kingdom
50 Hampshire Street, 5th Floor, Cambridge, MA 02139, United States

Notices
Knowledge and best practice in this field are constantly changing. As new research and experience broaden
our understanding, changes in research methods, professional practices, or medical treatment may become
necessary.

Practitioners and researchers must always rely on their own experience and knowledge in evaluating and using
any information, methods, compounds, or experiments described herein. In using such information or methods
they should be mindful of their own safety and the safety of others, including parties for whom they have a
professional responsibility.

To the fullest extent of the law, neither the Publisher nor the authors, contributors, or editors, assume any
liability for any injury and/or damage to persons or property as a matter of products liability, negligence or
otherwise, or from any use or operation of any methods, products, instructions, or ideas contained in the
material herein.

British Library Cataloguing-in-Publication Data
A catalogue record for this book is available from the British Library

Library of Congress Cataloging-in-Publication Data
A catalog record for this book is available from the Library of Congress

ISBN: 978-0-12-812209-9

For Information on all Elsevier publications
visit our website at https://www.elsevier.com/books-and-journals

Working together
to grow libraries in
developing countries

www.elsevier.com • www.bookaid.org

Publisher: Joseph Hayton
Acquisition Editor: Kathryn Morrissey
Editorial Project Manager: Amy Clark
Production Project Manager: Vijayaraj Purushothaman
Designer: Dayle G. Johnson and Matthew Limbert

Typeset by MPS Limited, Chennai, India

DEDICATION

For my mentors, past and present.

CONTENTS

When Dr. Hickey asked me to write a foreword to his monograph *Inhaled Pharmaceutical Product Development Perspectives*, my mind flashed back to the enthusiastic young college professor from the University of Illinois at Chicago that I first met more than 25 years ago and his enormous accomplishments since. Dr. Hickey has devoted his career of more than 35 years as Professor of Pharmacy to all aspects of the science and the development of pharmaceutical aerosols. He holds 17 patents, has published over 160 journal articles, and edited numerous books on the subject. In addition, he founded two pharmaceutical companies! He has worked many years as a volunteer member or chairman of USP Aerosol Expert Committees. In writing this monograph, Dr. Hickey uses his considerable experience and expertise to provide perspectives on five critical areas of inhaled pharmaceutical product development as an aid for those entering the field. The monograph largely covers the "Modern Era" of pharmaceutical aerosols that began in April 1955, when Susie asked her father, "Daddy, why can't they put my asthma medicine in a spray can like they do hair spray?" Susie's wise father, Dr. George Maison MD, President of Riker Laboratories, Inc., a subsidiary of the Rexall Drug Company, sought an answer to the question.

Dr. George Maison asked Irving Porush, lead chemist in Riker's three person Pharm. Dev. Lab located in Rexall's HQ building (just down the hall from helpful people in Rexall's Hair Spray lab) to look into answering Susie's question. Irv created two bronchodilator formulations: drug dissolved in a 50%v/v ethanol—chlorofluorocarbon propellant mix, packaged in plastic coated glass vials, sealed with Meshberg 50-μL metered dose valves and equipped with polyethylene mouthpieces. The FDA accepted NDAs for Medihaler Epi (OTC) and Medihaler Iso (Rx) on February 23, 1956. They were approved on March 9, 1956 and were on the market before the end of the month!

While helping out in Pharm. Dev. in August of 1956, I found that micronized bronchodilators could be dispersed in propellant using sorbitan trioleate as a dispersant. Spirometric measurements proved that micronized bronchodilators in suspension were far more effective than their counterparts in solution due to smaller delivered particle size. After safety studies, Riker submitted amendments to the NDAs to include the suspensions. Suspension versions of Medihaler Epi and of Medihaler Iso, packaged in 200 dose stainless steel vials equipped with 50-mcL metered dose valves, were marketed in the Fall of 1957. At that time "Proof of Safety" was the only requirement for an NDA approval. The amended NDA is less than 2 cm thick!

The regulatory climate changed dramatically on October 10, 1962 when President Kennedy signed into law the Kefauver–Harris amendment to the "Federal Food, Drug, and Cosmetic Act." Now, "proof of *effectiveness* and *safety*" are required before an NDA may be approved. These changes really are appropriate. However, sometimes the increased data required for NDA approval seems to be over the top! For example, the NDA approved in August 1996 for the change of propellant from CFCs to HFA-134a used in the Albuterol inhaler consisted of 193 volumes 400 pages each!

Dr. Hickey's monograph *Inhaled Pharmaceutical Product Development Perspectives* nicely encapsulates in six chapters the important aspects that one encounters in the development of new inhalers and the new formulations required to incorporate new or old drugs in them. Inhaled pharmaceuticals is a truly exciting and rewarding field in which to work. I wish Dr. Hickey's monograph had been available when I entered the field 60 years ago.

Charles G. Thiel
Retired, Division Scientist, 3M Drug Delivery Systems

This short overview of the practical considerations associated with inhaled pharmaceutical product development is intended to serve as a guide to the field and point of entry to the wealth of literature that has appeared largely in the last 25 years. The adoption of novel technologies or methods to facilitate inhaled drug therapy and new research findings should be considered in the context of the primary literature and the foundation of historical precedent. It is the intention of this monograph to outline methods and procedures while also commenting on product development strategies and their suitability to meet regulatory expectations. The value of this exposition is in identifying the opportunities and challenges and framing them with relevant reference material without complicating the higher level observations with the details that can be found in comprehensive texts covering elements of this topic. Informing those interested in the subject of standards against which to measure their expectations should serve to direct them to the literature that will address their specific questions.

An overview of analytical technologies, the proprietary landscape, development strategy, clinical and regulatory perspectives is given. By covering each of these areas, a framework is presented to guide those considering inhaled pharmaceutical product development. It is the author's objective to fill the gap between the aspiration of scientists invested in new technology development and the reality of the requirements that must be met for any new product to be commercialized. Focusing on underlying scientific and technical principles known to be acceptable from the current technical and regulatory perspective, this monograph should remain useful as a high-level guide to inhaled product development.

Anthony J. Hickey
Research Triangle Park, NC, United States
June 2017

ACKNOWLEDGMENTS

I am grateful to my professional colleagues, staff, postdoctoral fellows, and students who have been and continue to be a source of knowledge and inspiration. If the need for brevity has resulted in any omissions, though unintentional, I hope that I did not overlook anything that would substantially alter the conclusions. To my friends and colleagues who have toiled ceaselessly to construct the fabric of understanding that we have of pharmaceutical aerosols and inhaled medications I extend my heartfelt thanks. Your thorough research and keen observations are the foundation from which the opinions expressed in this monograph emerged.

Any project of this nature is a team effort regardless of authorship. I am extremely grateful for the editorial suggestions of Gerald Pollard and his timely review of all materials for the book. I also thank Amy Clark and Brian Thomas for their assistance in initiating the project and for editorial oversight on behalf of Elsevier and RTI International, respectively. Without their involvement, I would not have been able to navigate the publishing process with such ease.

CHAPTER *1*

Introduction—Historical Perspective

The use of smokes and mist aerosols for the treatment of disease is noted in many cultures from the beginning of recorded history. It is likely that inhaled materials have been therapeutic tools since time immemorial. Possibly the first recorded use of aerosols containing drugs is that of smokes of stramonium alkaloids in the Indus Valley thousands of years ago.[1] Attempts to formalize aerosol delivery paralleled discoveries in pharmacology since the Enlightenment period of the 18th century.[2] The inhalers and pharmaceutical aerosol dosage forms with which we are now familiar were developed in the second half of the last century starting with the pressurized metered dose inhaler in 1956 and progressing through a variety of dry powder inhalers in the 1960s and 1970s.[3] Nebulizer therapy had been readily available since the 19th century, but the device design and combination with known pharmacological agents demonstrated in controlled clinical trials occurred formally in the latter half of the 20th century.

This volume is structured in terms of general product development activities, which follow a pattern dictated by the process of bringing new drugs to market to address the needs of disease management. Fig. 1.1 illustrates the conventional image of the winnowing of drugs though the development process to ultimately gain regulatory approval and a path to commercialization.[4] As is evident, and frequently the subject of literature commentary, attrition during this process is significant and contributes to concerns about the overall efficiency, in both time and expense, of developing pharmaceuticals.[5]

Fig. 1.2 outlines the steps in product development as a guide to later discussion. The preclinical absorption, distribution, metabolism, and excretion; toxicology; and clinical studies will not be described in detail, as these activities only differ from conventional study designs in the route of administration. Chemistry, manufacturing, and controls to inform the regulatory considerations leading to clinical trials will be the major focus.

Inhaled Pharmaceutical Product Development Perspectives. DOI: https://doi.org/10.1016/B978-0-12-812209-9.00001-4

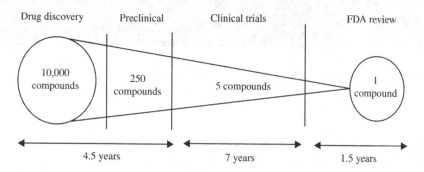

Figure 1.1 Attrition in the drug development pipeline. Modified from PhRMA Foundation (http://www.pharma-foundation.org).

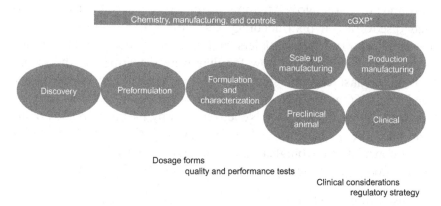

Figure 1.2 Drug development activities.

INHALED PRODUCTS

Inhaled therapy has been used largely for pulmonary disease. However, in response to the difficulty that the biotechnology industry experienced in delivering macromolecules, some emphasis has been placed on the lungs as a route of administration for systemically acting drugs.[6]

Asthma

Asthma has been a significant target for inhaled therapy, as it is a pulmonary airways disease. Since the mid-1950s when 3M Pharmaceuticals developed the first pressurized metered dose inhaler, asthma has been the major beneficiary of new drug product development.[2] The many research and clinical findings over the years of aerosol treatment of asthma have led to adjustments in its management.

Reviews of pharmacological therapy for asthma are numerous.[7,8] Initially, nonspecific β-adrenergic agonists were employed. They had adverse effects on the heart. Eventually, β$_2$-adrenergic agonists (BAs) were identified as having specificity suitable for bronchodilatation without adverse effects. Corticosteroids were delivered to treat the underlying inflammation, and anticholinergics [muscarinic antagonists (MAs)] were employed to achieve bronchodilatation through the parasympathetic nervous system. The relatively short residence time of the original BAs and MAs, which was suitable for relief of an acute exacerbation, was inconvenient for maintenance therapy to control the disease. Consequently, long-acting BAs and MAs (LABAs and LAMAs) were developed. Recently, the combination of some of these agents into single dosage forms has helped in suppressing the symptoms (LABAs and LAMAs) and treating the underlying cause of disease (corticosteroids) to bring about near normal airways and thereby allow management through routine treatment.

Chronic Obstructive Pulmonary Disease

Chronic obstructive pulmonary disease most frequently results from a history of smoking and involves characteristic impediment to gaseous exchange through local inflammation, airway remodeling, airway narrowing, and mucus thickening, often with infection.[9,10] Targeting the cholinergic receptors of the central airways with an LAMA was thought to be a practical approach to relieving symptoms,[11] so the first approved inhaled product was in this category, and subsequently combination products of LABAs and LAMAs were approved.

Cystic Fibrosis

Cystic fibrosis is a complex disease, the underlying cause of which is a genetic aberration in the ability to transport chloride ions.[12,13] This gives rise to thick mucus, which is not cleared efficiently by the mucociliary mechanism and forms a platform for microbial growth. The immune response to infection involves white blood cell (WBC) infiltration to the airways. These WBCs, at death, release deoxyribonucleic acid (DNA) that contributes to further thickening of the mucus. Aerosol treatment delivers an enzyme to cleave the DNA, thereby reducing viscosity, and antibiotics to treat the major infection with

Pseudomonas aeruginosa.[14] A range of mucolytics can be delivered that help in the clearance of mucus and improve lung function.

Diabetes

The benefit of using the lungs as a route for the delivery of systemi-cally acting agents was first embodied in a drug product in the early 1960s with an approved aerosol therapy for migraine headaches.[15] After a 30-year hiatus, interest in this route returned in the early 1990s when work began on the delivery of the protein insulin for the treat-ment of diabetes. After over 15 years of work, aerosol insulin was finally approved and commercialized in 2006.[16,17]

SCIENTIFIC ADVANCES

Progress has occurred in many areas of aerosol delivery. The founda-tional elements of pharmaceutical inhalation aerosol technology dis-cussed here include formulation, inhaler design, sampling, and characterization[18] and are treated in greater detail in later chapters.

Raoult's and Dalton's Laws

The behavior of propellant-based pharmaceutical aerosol delivery systems can be approximated by the application of Raoult's and Dalton's laws on the behavior of volatile liquids. Raoult's law esti-mates the vapor pressure of the final product based on the fraction of each of the components (propellant and cosolvents) with respect to the overall vapor pressure of the product.[18] Liquefied propellant offers the advantage that the vapor pressure supporting product per-formance does not diminish provided sufficient propellant is retained to support equilibrium with the atmosphere. Therefore, Raoult's law indicates that the partial vapor pressure (P_a) of the propellant is equivalent to the mole fraction (propellant, n_a, with respect to total including other components, $n_a + n_b$) multiplied by pure vapor pres-sure $P_a°$ (1.1). Likewise, the partial vapor pressure of the other com-ponents (P_b) is equivalent to the mole fraction (other components, n_b, with respect to total including propellant, $n_a + n_b$) (1.2). According to Dalton's law, the total vapor pressure of the product (P') is the sum of all partial vapor pressures (1.3), leading to the following expressions:

$$P_a = n_a/(n_a + n_b) \cdot P_a°   \tag{1.1}$$

$$P_b = n_b/(n_a + n_b) \cdot P_b^{\circ} \qquad (1.2)$$

$$P' = P_a + P_b \qquad (1.3)$$

Forces of Interaction in Dry Powders

Drug particles are incorporated into dry powder formulations intended to deliver drugs to the lungs. Consideration of the forces of interaction between particles is necessary to address the formulation and device interactions required to disperse drug for inhalation. The major forces of interaction that bind drug and excipient particles together are van der Waals, electrostatic, capillary, and mechanical interlocking.[19–21] The first three are subject to description by fundamental physicochemical properties as shown in Table 1.1. The fourth, mechanical interlocking, is difficult to characterize as, to predict behavior, control of surface rugosity must be assumed. At this point, surface rugosity, and by inference controlled mechanical interlocking, has been postulated as a means of controlling powder behavior but cannot be calculated from first principles.[22]

Aqueous Spray Formation

The dispersion of water droplets is governed by the characteristic force with which the liquid is pushed or drawn through an orifice jet and the forces of attraction of water molecules that must be overcome to create new surface, most notably hydrogen bonding. The dispersion of aqueous droplets has been described in detail. The atomization process can be divided into bulk fluid breakup and droplet formation.[24] It should be noted that, for water, viscous forces play a small role compared to surface tension. Exceeding the point at which aerodynamic forces acting to disrupt liquid in air are equal to the surface tension binding the

Table 1.1 Definitions of Forces of Particle Interaction		
Force	**Equation**	**Definitions**
Van der Waals	$\dfrac{AD_{12}}{12h^2}$	A is the Hamaker constant; $D_{12} = d_1 d_2/(d_1 + d_2)$, where d_1 and d_2 are particle diameters; h is the shortest distance between the particles[23]
Electrostatic	$\dfrac{K_q q_1 q_2}{h^2}$	q_1 and q_2 are the charges on the particles; h is the separation distance between the adhering particles; K_q electrostatic force constant[23]
Capillary	$2\pi\sigma D_{12}$	σ is surface tension; D_{12} same as above[23]
Modified from Hickey, A., Pharmaceutical inhalation aerosol powder dispersion—an unbalancing act. Am Pharm Rev, 2003. 6: p. 106–110.		

droplet together leads to droplet breakup and is defined by the critical Weber number[25,26]

$$We_{crit} = 8/C_D \qquad (1.4)$$

where C_D is the drag coefficient (~ 0.45 for turbulent conditions)[24] and

$$We = \left(\rho_A U_R^2 D\right)/\sigma \qquad (1.5)$$

where ρ_A is air density, U_R is the relative velocity between the droplet and the airstream, D is the diameter of the droplet, and σ is the surface tension.

Given these conditions, if the We exceeds 18, the droplet will disintegrate. This is the starting point to consider the behavior of nebulized droplets, but further consideration must be given to empirical and semi-empirical considerations to predict droplet behavior during atomization. One approach reduces to the following expression for water at $20°C$[27]:

$$D = \left(4998/U_R\right) + 904,411 \times \left(Q_l/Q_a\right)^{1.5} \qquad (1.6)$$

where D is the diameter of the droplets, U_R is the relative velocity between gas and liquid, and Q_l and Q_a are the flow rates of liquid and gas, respectively.

TECHNICAL ADVANCES

Advances during the period of modern inhaler design have been very broad, but few of the most progressive technologies have become products. Industry drivers of new development have been the most successful. The phase-out and elimination of chlorofluorocarbon propellants through the 1990s and the first decade of the new millennium in response to the observation of their involvement in ozone depletion was an important impetus.[28] The slowing of the biotechnology revolution of the late 1980s and early 1990s due to the lack of formal understanding of barriers to macromolecular drug delivery through which the option of aerosol delivery became a major consideration was a second and potentially more important influence from 1990 to 2010.

Pressurized Metered Dose Inhalers
Today's devices are similar in appearance to the originals over 60 years ago,[29] but appearance is deceiving. The most significant of many

developments was the phase-out of chlorofluorocarbon propellants as part of the Montreal Agreement on ozone depleting chemicals. The resultant industry thrust to reformulate in new propellants led to innovation in formulation, canister, valve, and actuator design and composition, in addition to the need to completely review filling equipment components and performance.

Dry Powder Inhalers

Dry powder inhaler technology has experienced major advances since the earliest devices were commercialized.[30–32] The basic principles of formulation, metering, and aerosol dispersion mechanism have largely remained unchanged. However, over time, there has been movement from primarily low efficiency unit-dose lactose blends delivered from inhalers with unique dispersion mechanisms to high efficiency multidose systems with more sophisticated delivery mechanisms. Moreover, the move from low dose to high dose has been accomplished by moving away from lactose blends to mostly pure drug systems.

Nebulizers

A wide range of nebulizers are available.[26] Most are air jet systems operating on Bernoulli or Venturi principle and consisting of three major components: a tube to convey high pressure air to converge on the tip of a tube immersed in solution/suspension of drug such that the latter is drawn up at high velocity and impinged on a baffle that collects the large droplets, which are returned to the reservoir. Small droplets are then delivered on the generating air supply to the patient. While the principles are the same for all jet nebulizers, some manufacturers have invested in optimizing the efficiency of delivery and maximizing the proportion of the dose delivered (minimizing the amount remaining in the nebulizer, the "dead volume").

Two other types are popular. The ultrasonic nebulizer uses a piezoelectric source to vibrate the drug solution/suspension thereby causing cavitation and buoyancy within the fluid and interference at the surface of the fluid, all of which gives rise to droplet formation. The vibrating mesh nebulizer also uses a piezo-electric aerosol generator, but it is built into an orifice plate through which drug solution/suspension is driven by the motion of the plate with respect to a small liquid reservoir. The advantage of these systems is that droplets can be generated in standing air and the volume flow can be adjusted to deliver the air to the patient while the operating parameters of the piezo-electrical system can be modulated to achieve the desired dose and particle size of aerosol output.

Other Technologies
Particle Design

Micronization is the classical method of particle production for use in any of the inhaler systems.[20,21,33] This is a controlled particle size reduction process in which large particles of drug are impinged on each other and the walls of the mill at high velocity, thereby reducing the size until particles can leave the mill as a function of their aerodynamic behavior. The control of this process relates to final particle size distribution but does not adequately control the bulk or surface properties of the final particles to allow them to be predictably used in a formulation for pulmonary drug delivery. The impact of the process on crystallinity, moisture content, hygroscopicity, shape, and size can significantly affect the ease with which the formulation can be prepared. Moreover, drug often must be mixed with a carrier lactose powder to be dispersed adequately as an aerosol, which limits the dose size considerably.

The use of spray drying to prepare particles of pure or almost pure drug has allowed reduction in forces of interaction, increasing the efficiency of dose delivery and thereby substantially increasing the highest dose of drug that can be administered from a dry powder inhaler. This facilitates the treatment of infectious disease, where drugs typically have modest potency and large doses are required.[34] The most prominent example is tobramycin.[35]

Other methods such as supercritical fluid and co- and countercurrent precipitation in propellant have all been used to prepare particles with unique physicochemical properties, but so far none have appeared in commercial products.[36–38]

Soft Mist

There is only one soft mist inhaler, the Respimat by Boehringer Ingelheim.[39] The principle is to meter the drug solution from a reservoir to two narrow inlet tubes engineered into a silicon wafer through which flow is impinged. High pressure applied to the fluid as it passes through the actuator tubes results in a breakup of the liquid stream into droplet sizes suitable for inhalation.

Counters and Patient Compliance Aids

A range of auxiliary devices have been used to increase patient compliance. The use of spacers and valved holding chambers was an early development intended to separate the actuation of pressurized metered dose inhalers and inhalation of the aerosol plume generated.[40] This reduced the need for coordination by the patient, particularly children and the elderly, and improved the efficiency of delivery and patient outcome. The impact was not only on efficacy but also on safety, particularly for steroids, where the reduction of oropharyngeal deposition led to fewer throat infections.

Counters were a relatively late development, coming almost 20 years after the first inhalers were developed and only being mandated in the last decade.[41] The advantage of counters on multidose devices is clear. The patient is aware of how many doses are left in the container and can be certain to obtain their medication in a timely fashion. A patient paying close attention can also tell whether they have missed a dose, and this will help with controlling acute exacerbations and maintaining a near normal life.

Many ventures have been made into device aids that report pulmonary function or generate data to be accessed by patient and physician to monitor drug use and therefore support guidance on compliance questions. None has been adopted for routine use in products to date. The technology to collect such data is readily available and is used in other health care strategies, notably in nutrition and exercise monitoring. Therefore, it seems likely that these features will be incorporated into products in the near future. One important point to bear in mind with respect to the speed of adoption of complex patient compliance technologies is the regulatory requirements that might be established and their impact on the speed of new product approval.

Micro- and Nanotechnology

Research findings have been reported on a wide range of micro- and nanoparticle technologies for aerosol delivery, but very few of these approaches have reached the clinic or commercialization.[42,43] Notable successes are the use of pure drug colloids for aqueous steroid formulations[44] and liposomal formulations of antimicrobial agents,[45,46] both intended for delivery by nebulizer.

REGULATORY ADVANCES

The major developments in regulatory consideration of aerosols relate to quality in manufacturing and product performance characterization in the context of chemistry and manufacturing controls.[47] As clarity occurs in the regulatory arena, the prospects of greater numbers of applications, speed of review, and approval will likely lead to more products being commercialized.

Orally Inhaled Drug Product Guidance

In 1998, the US Food and Drug Administration (FDA) first promulgated its draft guidance for developing pressurized metered dose inhalers and dry powder inhalers.[47] Notably, this document has not reached the stage of becoming formal guidance. Nevertheless, it has served the purpose for almost 20 years. As science and technology have progressed, this document has stood the test of time remarkably well; and while there may be ways in which it could be updated to reflect current knowledge, it remains a valuable tool for developing new products and contemplating generic strategies.

Quality by Design Guidance

At the beginning of this century, staff at the FDA and the pharmaceutical industry turned their attention to the approach to drug manufacturing.[48] Broadly speaking, and beyond the narrow scope of pharmaceutical aerosols, the manufacturing unit operations were largely batch processes at that time. The batch process strategy involved setting operating conditions at the beginning and controlling the quality of the product through testing of critical properties at the end. Products that did not meet specification were discarded. The need to redefine the process specifications was driven by the cost of out-of-specification observations being made.

Other industries have for decades used continuous processing, statistical process control, in-line monitoring control, and risk management strategies to manufacture products, and the time seemed right for the same principles to be adopted as a matter of regulatory mandate for pharmaceutical products.

Orally Inhaled Generic Drug Product Guidance

In the last 5 years, the FDA has released many generic guidance documents, which will be discussed in Chapter 5. They are directed at specific products and are clearly helpful for those wishing to gain regulatory approval for bioequivalent drugs. However, regulatory strategy is not the only consideration in generic product development. In some cases, it may be difficult to match innovator technology exactly because of intellectual property or trade secrets, which leave gaps in technical understanding that may be essential to the production of equivalent products.

CLINICAL ADVANCES

The basis of our understanding of deposition of particles in the lungs evolved from the beginning to the middle of the last century in the fields of environmental health and occupational medicine. The work of Findeison and Landahl was the foundation of a growing understanding of the importance of particle size and other factors in the deposition and clearance of aerosols.[49−51] By 1966, the Task Group on Lung Dynamics of the International Commission on Radiation Protection had collated everything that was known at the time to give insight into the deposition of aerosols in the lungs.[52] A combination of experimental and theoretical data gave a clear indication that regional deposition was dictated by aerodynamic particle size distribution. While more experimental and theoretical data have been generated in the interim, the basic principle established 50 years ago that the target aerodynamic size distribution for lung delivery is 1−5 μm remains true. The objective of all technologies is to deliver droplets or particles in this size range.[53] Understanding lung deposition and subsequent clearance is a foundational step to potentially predicting therapeutic outcomes following aerosol therapy.

CONCLUSION

Following millennia of anecdotal use of pharmaceutical aerosol therapy, modern formularies are replete with well-designed, highly engineered inhaled drug delivery systems of a quality sufficient to meet the performance standards required of modern products. The ability to harness understanding from the latest scientific inquiry in the service of the industrial design of pharmaceutical aerosol products that can meet the demands of regulatory scrutiny is the objective of those working in the field. The convergence of the science and technology underpinning drug manufacture, formulation, metering, inhaler design, and assembly into the final product with regulatory expectations of relevant quality and performance measures based on critical attributes is the foundation for new and generic product development. The following chapters capture the key steps in inhaled pharmaceutical product development.

REFERENCES

1. Jackson M. "Divine stramonium": the rise and fall of smoking for asthma. *Med Hist.* 2010;54:171–194.

2. Stein S, Thiel C. The history of therapeutic aerosols: a chronological review. *J Aerosol Med Pulmonary Drug Delivery.* 2017;30:20–41.

3. Sanders M. Pulmonary drug delivery: an historical overview. In: Smyth H, Hickey A, eds. *Controlled Pulmonary Drug Delivery.* New York: CRS-Springer; 2011:51–73.

4. http://www.pharmafoundation.org.

5. Scannel J, et al. Diagnosing the decline in pharmaceutical industry R&D efficiency. *Nat Rev: Drug Discovery.* 2012;11:191–200.

6. Adjei A, Gupta P. *Inhalation Delivery of Therapeutic Peptides and Proteins.* New York: Marcel Dekker; 1997.

7. Crooks P, Al-Ghananeem A. Drug targeting to the lung: chemical and biochemical considerations. In: Hickey A, ed. *Pharmaceutical Inhalation Aerosol Technology.* New York: Marcel Dekker; 2004:89–154.

8. Hickey A. Pulmonary drug delivery: pharmaceutical chemistry and aerosol technology. In: Wang B, Hu L, Siahaan T, eds. *Drug Delivery, Principles and Applications,* Second ed. New York: John Wiley and Sons; 2016:186–206.

9. Balkissoon R, et al. Chronic obstructive pulmonary disease: a concise review. *Med Clin North Am.* 2011;95:1125–1141.

10. Barnes P. Chronic obstructive pulmonary disease. *N Engl J Med.* 2000;343:269–280.

11. Chan J, et al. Advances in device and formulation. technologies for pulmonary drug delivery. *AAPS PharmSciTech.* 2014;15:882–897.

12. Rowe S, Miller S, Sorscher E. Cystic fibrosis. *N Engl J Med.* 2005;352:1992–2001.

13. Cutting G. Cystic fibrosis genetics: from molecular understanding to clinical application. *Nat Rev Genet.* 2015;16:45–56.

14. Garcia-Contreras L, Hickey A. Pharmaceutical and biotechnological aerosols for cystic fibrosis therapy. *Adv Drug Delivery Rev.* 2002;54:1491–1504.

15. Hickey A. Back to the future: inhaled drug products. *J Pharm Sci.* 2013;102:1165–1172.

16. Profit L. Exubera (inhaled insulin): an evidence-based review of its effectiveness in the management of diabetes. *Core Evidence.* 2005;1:89–101.

17. Barnett A. Exubera inhaled insulin: a review. *Int J Clin Pract.* 2004;58:394–401.

18. Hickey A, Mansour H. Delivery of drugs by the pulmonary route. In: Florence A, Siepmann J, eds. *Modern Pharmaceutics.* New York: Taylor and Francis; 2009:191–219.

19. Hickey A. Pharmaceutical inhalation aerosol powder dispersion—an unbalancing act. *Am Pharm Rev.* 2003;6:106–110.

20. Hickey A. In: Merkus H, ed. *Fundamentals of dry powder inhaler technology.* New York: AAPS-Springer; 2017.

21. Hickey A. *Complexity in pharmaceutical powders for inhalation: a perspective.* KONA, 2017: Published ahead of print.

22. Chew N, Chan H. Use of corrugated particles to enhance powder aerosol performance. *Pharm Res.* 2001;18:1570–1577.

23. Dunbar C, Holzner P, Hickey A. Dispersion and characterization of pharmaceutical dry powder aerosols. *KONA.* 1998;167:433–441.

24. Hinds W. *Aerosol Technology Properties, Behavior and Measurement of Airborne Particles.* Second ed New York: John Wiley and Sons; 1999.

25. Lefebvre A. *Atomization and Sprays.* New York: Hemisphere Publishing Corporation (Taylor and Francis Group); 1989.

26. Niven R, Hickey A. Atomization and nebulizers. In: Hickey A, ed. *Inhalation Aerosols, Physical and Biological Basis for Therapy.* New York: Informa Healthcare; 2007:253–283.

27. Nukiyama, S. and Y. Tanasawa, *An experiment on the atomization of liquid by means of an air stream.* Trans Soc Mech Eng (Japan), 1939. 5 (summary section): p. S15–17.

28. Molina M, Rowland F. Stratospheric sink for chlorofluoromethanes: chlorine atom-catalysed destruction of ozone. *Nat Rev Genet.* 1974;249:810–812.

29. Purewal T, Grant D. *Metered Dose Inhaler Technology.* Deerfield, IL: Interpharm Press; 1998.

30. Boer AD, et al. Dry powder inhalation: past, present and future. *Expert Opin Drug Delivery.* 2017;14:499–512.

31. Atkins P. Dry powder inhalers: an overview. *Respir Care.* 2005;50:1304–1312.

32. Frijlink H, Boer Ad. Dry powder inhalers for pulmonary drug delivery. *Expert Opin Drug Delivery.* 2004;1:67–86.

33. Hickey A. Lung deposition and clearance: what can be learned from inhalation toxicology and industrial hygiene? *Aerosol Sci Technol.* 1993;18:290–304.

34. Edwards D, et al. Large porous particles for pulmonary drug delivery. *Science.* 1997;276:1868–1872.

35. Geller D, Weers J, Heuerding S. Development of an inhaled dry-powder formulation of tibramycin using pulmosphere technology. *J Aerosol Med Pulmonary Drug Delivery.* 2011;24:175–182.

36. Oort MV, Sacchetti M. Spray-drying and supercritical fluid particle generation techniques. In: Hickey A, ed. *Inhalation Aerosols, Physical and Biological Basis for Therapy.* Second ed. New York: Informa HealthCare; 2007:307–346.

37. Matteucci M, et al. Design of potent amorphous drug nanoparticles for rapid generation of highly supersaturated media. *Mol Pharmaceutics.* 2007;4:782–793.

38. Wan J, et al. Microfluidic generation of droplets with a high loading of nanoparticles. *Langmuir.* 2012;28:13143–13148.

39. Lyseng-Williamson K, Keating G. Tiotropium Respimat Soft Mist inhaler: a guide to its use in chronic obstructive pulmonary disease (COPD) in the EU. *Drug Ther Perspect.* 2014.

40. Mitchell J, Dolovich M. Clinically relevant test methods to establish in vitro equivalence for spacers and valved holding chambers used with pressurized metered dose inhalers (pMDIs). *J Aerosol Med Pulmonary Drug Delivery.* 25: p. 217–242.

41. US Food and Drug Administration, *Integration of dose-counting mechanisms into MDI drug products.* 2003.

42. Gupta P, Hickey A. Contemporary approaches in aerosolized drug delivery to the lung. *J Controlled Release.* 1991;17:129–148.

43. Kuzmov A, Minko T. Nanotechnology approaches for inhalation treatment of lung diseases. *J Controlled Release.* 2015;219:500–518.

44. Skoner D, et al. Clinical use of nebulized budesonide inhalation suspension in a child with asthma. *J Allergy Clin Immunol.* 1999;104:210–214.

45. Rose S, et al. Delivery of aerosolized liposomal amikacin as a novel approach for the treatment of nontuberculous mycobacteria in an experimental model of pulmonary infection. *PLoS One.* 2014;9:e108703.

46. Elborn J. Ciprofloxacin dry powder inhaler in cystic fibrosis. *BMJ Open Respir Res.* 2015;3: e000125.

47. US Food and Drug Administration, *Draft guidance for the industry, metered dose inhaler (MDI) and dry powder inhaler (DPI) chemistry manufacturing and controls documentation.* 1998.

48. US Food and Drug Administration, *Guidance for industry: Q8(R2) pharmaceutical development.* 2009.

49. Findeisen W. Uber das absetzen kleiner, in der luft suspendierten teilchen inder menschlichen lunge bei der atmung. *Arch Ges Physiol.* 1935;236:367–379.

50. Landahl H. On the removal of airborne droplets by the human respiratory tract. I: The lung. *Bull Math Biol.* 1950;12:43–56.

51. Landahl H. On the removal of airborne droplets by the human respiratory tract. II: The nasal passages. *Bull Math Biophys.* 1950;12:161–165.

52. Task Group on Lung Dynamics. Deposition and retention models for internal dosimetry of the human respiratory tract. A report prepared by the Task Group on Lung Dynamics for Committee II of the International Commission on Radiological Protection. *Health Phys.* 1966;12:173–207.

53. Hickey A. Summary of common approaches to pharmaceutical aerosol administration. In: Hickey A, ed. *Pharmaceutical Inhalation Aerosol Technology.* Second ed. New York: Marcel Dekker; 2004:385–421.

CHAPTER 2

Dosage Forms

Inhaled dosage forms fall into three major categories and a small number of individual systems. Each of the conventional dosage forms—metered dose inhalers, dry powder inhalers (DPIs), and nebulizers—is discussed in terms of the formulation, the metering system, and the aerosol dispersion mechanism or actuator. To illustrate current progress, alternative dosage forms, delivery systems, and formulations are discussed, indicating the advantages that these systems offer in the delivery of drugs to the lungs.

METERED DOSE INHALERS

Formulation

Metered dose inhalers are frequently termed pressurized metered dose inhalers (pMDIs) as the drug is formulated in high vapor pressure propellant that gives rise to high pressures within the canister. The propellant flash evaporates at atmospheric pressure, and the ability to harness this property is the basis for pMDI performance. In addition to the propulsion and droplet generation that can be achieved through this evaporative process, the propellant serves as the medium in which to disperse drug in either solution or suspension. Currently, the prominent propellants HFA 134a and 227 are combined with the drug, and the interaction is mediated by the use of the cosolvent ethanol and the surfactant oleic acid.[1,2]

Propellants are named according to their chemical composition.[3] The number on the right (4 in 134a) represents the number of fluorine atoms, the middle number (3 in 134a) represents one more than the number of hydrogen atoms (H + 1), and the number on the left (1 in 134a) represents one less than the number of carbon atoms (C − 1). The lowercase letter represents the first asymmetric isomer of the structure. For example, propellant 134a has the general structure $C_2H_2F_4$ and propellant 227 has the general structure C_3HF_7.

Solution formulations involve the homogeneous molecular dispersion of drug in propellant alone or facilitated by other additives.[4]

Inhaled Pharmaceutical Product Development Perspectives. DOI: https://doi.org/10.1016/B978-0-12-812209-9.00002-6

Suspension formulations require the use of other additives to stabilize the micronized drug and support uniform dose and aerodynamic particle size properties. Fig. 2.1 depicts the interactions that might occur for suspension in propellant, with cosolvent (ethanol, EtOH), and with cosolvent and surfactant (e.g., oleic acid). The association of the cosolvent or the surfactant at the surface of the drug suspension helps reduce the tendency for the particles to associate and thereby allows them to remain suspended within the nonpolar environment of the propellant. The objective is to allow for controlled flocculation on standing that can be readily deflocculated by shaking, allowing delivery of particles in near primary sizes.[4,5]

Metering System and Canister

The metering valve was the major advance that facilitated the development of pMDI technology.[6] The ability to sample small quantities of drug suspended or dissolved in small volumes of propellant-based formulation was an innovation that allowed the delivery of potent pharmacological agents for local action in the lungs. Fig. 2.2 illustrates the reservoir canister (the cylindrical insert in the left panel, photograph in the upper right panel), the valve (just below the canister on the left, photographs of two examples in the upper right panel), and the valve stem and actuator just below the valve. The lower right panel is a

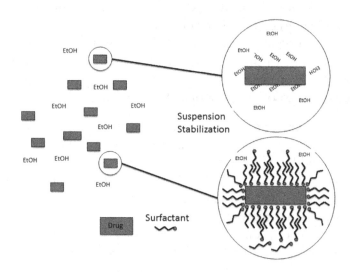

Figure 2.1 Formulation in propellant alone, with cosolvent (EtOH), and with cosolvent and surfactant. The presence of two or more drugs increases the physicochemical complexity.

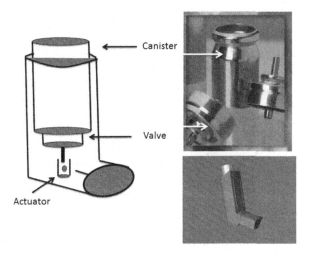

Figure 2.2 Metered dose inhaler and components.

photograph of the whole product. Valves are complex constructions that achieve two functions concurrently, closing the dosing chamber component of the valve to the reservoir in the canister and opening the dosing chamber through the valve stem to atmosphere. To achieve this and also to seal the valve to the canister, several gaskets and O-rings of different elastomer composition (e.g., rubber) and springs are required. Each of the component materials must be evaluated for compatibility with the product and for the gaskets and O-rings; extractables and leachables that might result in poor product performance or safety concerns must be measured.[7]

The canister is most frequently aluminum, but the use of the cosolvent ethanol and occasionally the capacity of the active pharmaceutical ingredient to associate with water mean that the canister has to be coated, usually with plastic, to prevent interactions.[8]

Actuator and Mouthpiece

Fig. 2.2 shows the actuator into which the valve and canister fit.[2] The valve stem sits on the orifice block, and when the canister is pressed the valve stem opens, allowing the metered volume to be discharged through the actuator orifice into the mouthpiece and to the patient, as illustrated in Fig. 2.3. While the valve was the most significant development in accurately and reproducibly metering propellant-based formulation, the actuator and mouthpiece were also important in that

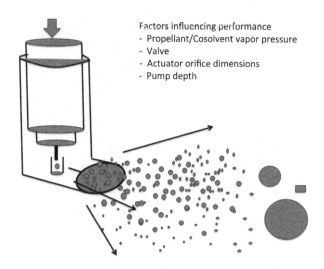

Factors influencing performance
- Propellant/Cosolvent vapor pressure
- Valve
- Actuator orifice dimensions
- Pump depth

Figure 2.3 Aerosol plume formation from a metered dose inhaler.

they contribute to the delivery of the drug by assisting in control of the particle size and plume geometry required to direct the dose to the patient.

Performance

Each component of the pMDI has a role in the final performance of the product. Fig. 2.4 shows two important considerations in the selection and use of formulations. Fig. 2.4A indicates the influence of vapor pressure on droplet size as the valve empties under the propulsive force of propellant evaporation.[1] It should be noted that because of the unique physicochemical properties of each propellant, evaporation proceeds at different rates, giving rise to different droplet size distributions. This makes matching of performance between propellants difficult to achieve. Fig. 2.4B depicts an idealized aerodynamic particle size distribution and the impact of Ostwald ripening and aggregation on suspension product performance.[9] Ostwald ripening is the phenomenon of dissolution of small particles and growth of large ones that occurs with temperature cycling and results in a shift towards larger particle size and narrowing of the distribution. Aggregation occurs when the suspension is insufficiently stabilized by the formulation; irreversible aggregation occurs such that larger particle constructs exist in the product, which usually results in a broadening of the distribution.[10] Fig. 2.5 illustrates the impact that such time-dependent phenomena

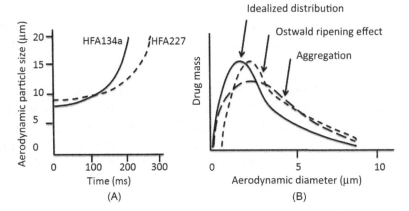

Figure 2.4 (A) Droplet formation on actuation and (B) aerodynamic particle size distributions illustrating the effects of instability. The impact of formulation variables on product performance is complicated significantly in drug combinations.

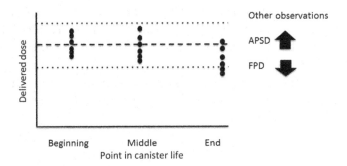

Figure 2.5 Effects of instability on delivered dose through the life of the canister. Where drug combinations are used, the performance of each drug must meet regulatory requirements for uniform and reproducible delivery.

may have on product stability. Anything that increases the aerodynamic particle size distribution will reduce the fine particle fraction (proportion of the dose entering the lungs) and ultimately lead to the product performing outside the desired specification.[11,12]

There may be other sources of instability in pMDI products that arise from the effectiveness of the valve in preventing entry of extraneous substances. Notably, ingress of moisture and silicone fluid into the product may affect performance.[13,14] Silicone is included as a lubricant in the valve but is not intended intentionally to enter the product.

Other

As experience was gained with the performance of pMDIs, two areas of improvement occurred with respect to patient compliance and adherence.

The first was the use of auxiliary spacer or valved holding chamber technologies that disconnected actuation of the device from inhalation.[15,16] In addition, most of these systems allowed for further evaporation to achieve smaller particle sizes. The reduction in velocity and collection of very large particles that occurs in these devices also limits the mouth and throat deposition.

The second area of improvement was the adoption of dose counters.[8,17,18] This is particularly important because it informs the patient when to order their next inhaler. Most pMDIs are overfilled; consequently, there are more doses than indicated on the dose counter. However, as the inhaler is depleted beyond the final dose, failure occurs precipitously at a critical volume of residual propellant. It would be unwise, especially for rescue medications such as short acting β2-adrenergic agonist bronchodilators, to allow the patient to use the inhaler for many doses after the counter indicates that the canister is empty, as underdosing could be a potentially fatal exacerbation.

DRY POWDER INHALERS

Formulation

Drug particles for DPIs are commonly milled to the appropriate size for lung delivery.[19] Lactose blends are the most popular formulations.[20] This strategy was adopted for the earliest products developed in the 1960s and 1970s.[21] The intent of preparing blends is to address specific limitations of inhaled drug therapy. Fig. 2.6 shows an electron micrograph of fine drug particles on coarse lactose carrier particles and a schematic of the intent to disperse the drug on the surface to allow ease of removal.[22] The objective of formulation is to achieve a balance of the forces of interaction that will stabilize the product through filling and storage but allow ease of dispersion when used by the patient.[23] The small quantities of drug usually delivered in these devices, most frequently 100 μg or less, cannot be filled readily and require a diluent in which they are homogeneously mixed. In addition,

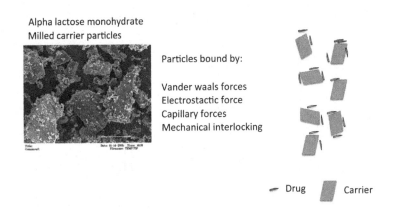

Alpha lactose monohydrate
Milled carrier particles

Particles bound by:

Vander waals forces
Electrostactic force
Capillary forces
Mechanical interlocking

Drug   Carrier

Figure 2.6 Example of dry powder drug carrier powders, and forces by which drug particles are bound to carrier particles.

the interparticulate forces between micronized particles can only be overcome by using enormous energy to impart dispersion forces. The inspiratory flow of the patient on which most DPIs deliver the drug is not sufficient to disperse micronized powder alone. Blending drug in low concentrations with lactose reduces the adhesive forces and results in lower cohesive forces that may be overcome to release drug onto the inspiratory airflow of the patient.[24,25]

Spray drying is an alternative approach to the preparation of dry powders that does not require subsequent blending with lactose and results in readily dispersible aerosol particles.[19]

Metering System

As DPI technology has evolved, a range of metering systems has been developed.[26] Fig. 2.7 shows schematically the various metering systems available to accommodate drug doses. Broadly, they fall into the categories of unit dose, multiunit dose, and reservoir systems. Unit dose metering systems were developed initially by adopting capsules from oral drug delivery. This approach was efficient in that the capsule could be disposed of after use and the inhaler was reusable, minimizing waste and rendering the metering system as simple as possible.

As new technologies appeared, the opportunity to help the patient by including sufficient doses in the inhaler to manage the disease for a reasonable time period without requiring continual disassembly and

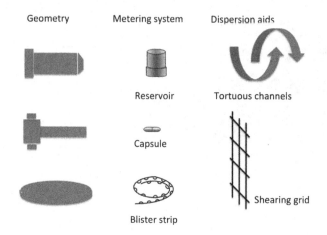

Figure 2.7 Dry powder devices, metering systems, and dispersion features.

reassembly of the device was addressed chronologically. The next system in development required metering the dose within the device by sampling from a reservoir of drug. The principle adopted might be considered a significantly scaled-down die, similar to that used in tableting, which, based on bulk density when filled, meters an accurate and reproducible dose. Unfortunately, this arrangement required considerable emphasis on storage stability due to the potential ingress of moisture that may affect the chemical and physical stability of the drug.[27]

An intermediate solution between unit dose and reservoir systems was the next step. The multiunit dose systems appear in many different embodiments, most notably blister disks and strips.[28] The intent is again to give the patient a drug supply sufficient to minimize manipulation of the device and to support disease management. A total of 30 or 60 doses that contain a month's supply of drug are common. The individual dose packaging is a major advantage of these metering systems, as it gives control over stability on storage with a well-understood packaging approach.

Actuator and Mouthpiece

All commercially available DPIs deliver the drug on the inspiratory flow of the patient. As a result, the interplay of the device geometry with the airflow plays a crucial role in the dispersion of drug and its delivery to the lungs.[29,30] Fig. 2.8 illustrates the use of design features, in this case a turbulence grid to assist in the separation of fine particles from the surface of the lactose particles delivering them on the inspiratory flow.[31]

NEBULIZERS

Formulation

All nebulizer formulations are aqueous based. Most are solutions of drug, but some are suspensions.[32] Occasionally cosolvents are employed, particularly ethanol, but this is rarely the case. Solution chemistry for nebulizers follows a similar pattern to that of parenteral formulation except there are a very small number of approved additives for use in inhaled products. Consequently, the major considerations for the preparation and stability of solutions for nebulization are pH and ionic strength. Nebulizer performance in terms of the droplet size distribution produced is also affected by surface tension and viscosity. The latter is only important where agents such as glycerol, propylene glycol, or polyethylene glycol have been added to significantly change the water properties. Suspension formulations are usually colloidal whether they consist of drug particles alone or lipid based systems such as liposomal small unilamellar vesicles.

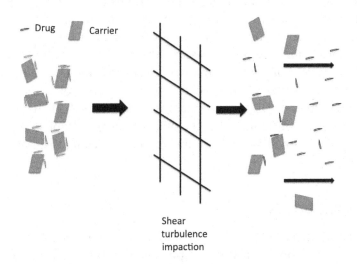

Figure 2.8 Dry powder aerosol dispersion from carrier based systems.

Metering System

Nebulizers are unique in that the nebulizer and the drug solution or suspension are not necessarily sold together and are not usually sold as a unit. Consequently, the metering system is usually not an integral part of the nebulizer. The dose is metered from a nebule into the reservoir of jet nebulizers, or to the orifice plate in the case of vibrating mesh systems.

Delivery Mechanism

The aerosol generation and delivery mechanism for nebulizers differs from any other inhaler in that all elements required for function are not integral to the device. Air jet nebulizers consist of a tube that dips into the reservoir and a tube, frequently coaxial, that directs air over the liquid feed tube to create a high velocity, low pressure region that draws liquid by Venturi or Bernoulli effect.[33,34] High velocity air is supplied by a separate pump. The solution or suspension is then dispersed on the airflow as droplets. Due to the hydrogen bonding of water, the spray consists of both large and small droplets. The small droplets are suitable for lung delivery. The large droplets impinge on a baffle and return to the reservoir. This recycling mechanism requires delivery of the solution or suspension as a steady-state aerosol over a period sufficient to deliver a therapeutic dose, which is frequently 15 to 30 minutes depending on the drug employed.

Vibrating mesh nebulizers have their origins in a method of producing calibration standards that has been available for 50 years, the vibrating orifice monodisperse aerosol generator (VOAG).[35] The VOAG had a single orifice in a piezo-electrical plate with closely controlled frequency of oscillation to produce droplets in a narrow size range. The evolution of this technology to a series of orifices in the plate, now renamed a mesh, resulted in a novel method of producing high density (mass/volume of air) clouds of aerosol droplets that could be delivered on small volumes of air, thereby reducing the overall time of dosing.[36–39]

OTHER PRODUCTS

Soft mist inhalers (SMIs) were developed as an aqueous based alternative to the propellant and dry powder hand-held aerosol delivery devices.[40] The most prominent SMI is Boehringer Ingelheim's

Respimat (Fig. 2.9).[38,40] The mechanism of delivery was originally published as the BINEB and consists of two channels in a silicon chip that impinge on liquid under pressure applied from the potential energy of a coiled spring.[41,42] The collision of the small liquid streams results in very small droplets containing drug in solution that can be inhaled in a manner similar to a pMDI. Other SMIs use a vibrating mesh with a mechanism similar to that used in vibrating mesh nebulizers, but with a liquid metering component from a reservoir to limit the dose to the vibrating mesh and thereby reduce the duration of delivery. Fig. 2.10 shows a concept for a hand-held vibrating mesh device (Aerohaler, from Aerogen). The Fox (Vectura) and EFlow (Pari) are based on vibrating mesh technology.[37]

The electrohydrodynamic spray mechanism of aerosol generation was also pursued for application in inhalers.[43] The principle of aerosol generation was to meter liquid through an orifice and form a Taylor cone.[44] This method involves the forced introduction of electrical charge to the surface of the emitted liquid, which increases its surface to accommodate the charged ions and ultimately forms droplets through thinning of the increasing surface.[44] This mechanism for producing small droplets has been used on a small scale for analytical methods such as mass spectroscopy and on a large scale for crop spraying. Its most prominent application in pharmaceutical drug delivery was for the treatment of lung cancer with doxorubicin.[43]

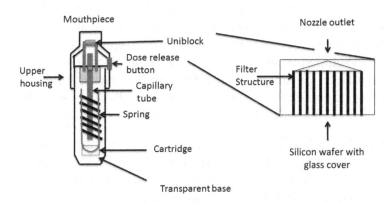

Figure 2.9 Respimat soft mist inhaler.

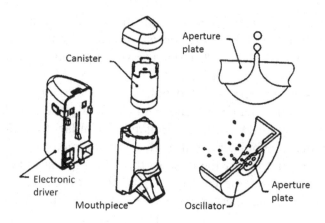

Figure 2.10 Vibrating-mesh hand-held configuration (Aerohaler from Aerogen).

SPECIALIZED FORMULATIONS

The inhalers described above use common formulation strategies, but alternative strategies have been developed for each of them. Lipid-based formulation is a notable new approach. Microemulsion preparations have been evaluated for MDIs to facilitate the formulation of proteins and peptides.[45–47] Solid lipid nanoparticles prepared by spray drying have been used in DPIs,[48] notably for tobramycin.[49] Liposomal formulations for nebulizer delivery have been developed, most notably for the delivery of high dose antibiotics such as ciprofloxacin and amikacin.[50,51]

MANUFACTURING

Each dosage form is manufactured by a series of unit operations, from active pharmaceutical ingredient manufacture to final packaging. In general, nebulizers are marketed separately from the drug. The drug may be in solution or suspension at the required dose or with the intention of dilution in saline, or in a dry form for reconstitution. Liquid formulations are frequently metered in a form-fill-seal process that presents the dose in a sealed container at the time of use.

Fig. 2.11 depicts the major processing steps in the manufacture of pMDIs[52] and DPIs.[53] In many respects, these processes are similar, with the exception that the high vapor pressure liquid propellant formulation requires filling at low temperature directly into the canister before valve placement and crimping, or at room temperature through

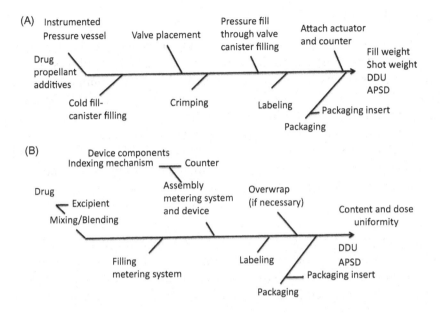

Figure 2.11 Manufacturing unit operations for (A) metered dose inhaler and (B) dry powder inhaler.

the valve. In both cases the propellant is dispensed in a manner that prevents evaporation. Dry powder, in contrast, can be handled at room temperature but requires specialized filling approaches to manage the low doses and metering systems used. In both cases, scale-up from bench manufacture to production requires considerable optimization to ensure the accuracy and reproducibility of dose delivery.

PERFORMANCE CRITERIA

In Chapter 3, in vitro methods of evaluating pharmaceutical aerosol products for their key performance criteria of delivered dose and aerodynamic particle size will be discussed.[54] These parameters are important as quality metrics that underpin regulatory considerations, which will be discussed in Chapter 5. They are also critical properties that link lung deposition, pharmacokinetics, and pharmacodynamics (i.e., therapeutic efficacy), which will be covered in Chapter 4.[55] The objective of these processes is to achieve the optimal combination of formulation, metering system, and mechanism of aerosol dispersion to accurately and reproducibly deliver drug to meet the therapeutic needs of the patient.

CONCLUSION

Several types of aerosol devices delivering a wide range of drugs are available. The most common are pMDIs, DPIs, and nebulizers. A few alternative systems have been developed, the most prominent from a therapeutic and commercial standpoint being the SMI. The systems may be conveniently considered in terms of their key components: formulation, metering system, and aerosol generation mechanism (actuator and mouthpiece). The metering valve and the pMDI were the major early innovations that allowed inhaled therapy to become commonplace. Early DPI and nebulizer technology was a logical extension of knowledge gained from oral and parenteral products. As new challenges were faced, particularly the phase-out and elimination of ozone depleting propellants, greater emphasis was placed on DPI and nebulizer technology. This coincided with the need for alternative routes and means of delivery of biotechnology products and the emergence of diseases that required higher dose delivery or complex combinations of drugs.

More inhaled drug products are available today than at any time in the past. Understanding of the efficient and reproducible delivery of drug to the biological target is the goal of pharmaceutical product development. Control of formulation and device specifications establishes the quality and performance of the inhaler product. The underlying physicochemical properties and aerosol physics limits the control that can be achieved. Minimizing the number of factors contributing to performance will lead to greater control of efficiency and reproducibility of drug delivery.

REFERENCES

1. Hickey A, Smyth H, Evans R. Aerosol generation from propellant-driven metered dose inhalers. In: Hickey A, ed. *Inhalation Aerosols, Physical and Biological Basis for Therapy.* Second ed. New York: Informa Healthcare; 2007:417–440.

2. Purewal T, Grant D. *Metered Dose Inhaler Technology.* Deerfield, IL: Interpharm Press; 1998.

3. Mcketta Jr. JJ. *Encyclopedia of Chemical Processing and Design: Volume 2—Additives to Alpha.* New York: Marcel Dekker; 1977.

4. Smyth H, Hickey A. Multimodal particle size distributions emitted from HFA-134a solution pressurized metered-dose inhalers. *AAPS PharmSciTech.* 2003;4:E38.

5. Hickey A, Dalby R, Byron P. Effects of surfactants on aerosol powders in suspension. Implications for airborne particle size. *Int J Pharm.* 1988;42:267–270.

6. Porush I, Thiel C, Young J. Pressurized pharmaceutical aerosols for inhalation therapy I. Physical testing methods. *J Pharm Sci.* 1960;49:70–72.

7. Norwood D, et al. Best practices for extractables and leachables in orally inhaled and nasal drug products: an overview of PQRI recommendations. *Pharm Res.* 2007;25:727–739.

8. Stein S, et al. Advances in metered dose inhaler technology: hardware development. *AAPS PharmSciTech.* 2014;15:326–338.

9. Myrdal P, Sheth P, Stein S. Advances in metered dose inhaler technology: formulation development. *AAPS PharmSciTech.* 2014;15:434–455.

10. Murnane D, Martin G, Marriott C. Investigations into the formulation of metered dose inhalers of salmeterol xinafoate and fluticasone propionate microcrystals. *Pharm Res.* 2008;25:2283–2291.

11. Ninbovorl J, Sawatdee S, Srichana T. Factors affecting the stability and performance of ipratropium bromide; fenoterol hydrobromide pressurized-metered dose inhalers. *AAPS PharmSciTech.* 2013;14:1294–1302.

12. Ivey J, Vehring R, Finlay W. Understanding pressurized metered dose inhaler performance. *Expert Opin Drug Delivery.* 2015;12:901–916.

13. Williams R, Hu C. Moisture uptake and its influence on pressurized metered dose inhalers. *Pharm Dev Tech.* 2000;5:153–162.

14. Fallon J, Peyron I, Hickey A. Effects of direct spiking of silicone oil into a model pMDI formulation. *Drug Dev Ind Pharm.* 2013;39:681–686.

15. Nagel M, et al. Performance of large- and small-volume valved holding chambers with a new combination long-term bronchodilator/anti-inflammatory formulation delivered by pressurized metered dose inhaler. *J Aerosol Med.* 2002;15:427–433.

16. Smyth HC, et al. The influence of formulation and spacer device on the in vitro performance of solution chlorofluorocarbon-free propellant-driven metered dose inhalers. *AAPS PharmSciTech.* 2004;5:E7.

17. Kaur I, Aggarwal B, Gogtay J. Integration of dose counters in pressurized metered-dose inhalers for patients with asthma and chronic obstructive pulmonary disease: review of evidence. *Expert Opin Drug Delivery.* 2015;12:1301–1310.

18. US Food and Drug Administration, Center for Drug Evaluation and Research (CDER). *Guidance for Industry—Integration of Dose-Counting Mechanism into Metered Dose Inhaler Products*; March 2003.

19. Hickey A. Complexity in pharmaceutical powders for inhalation: a perspective. *KONA Powder Part* J, 2017; (35): Advanced pub J-stage.

20. Hickey A, et al. Physical characterization of component particles included in dry powder inhalers I. Strategy review and static characteristics. *J Pharm Sci.* 2007;96:1282–1301.

21. Hickey A. Summary of common approaches to pharmaceutical aerosol administration. In: Hickey A, Ed. *Pharmaceutical Inhalation Aerosol Technology*, Second ed. New York: Marcel Dekker, Inc.; 2004; 385–421.

22. Hickey A, et al. Physical characterization of component particles included in dry powder inhalers II. Dynamic characterization. *J Pharm Sci.* 2007;96:1302–1319.

23. Hickey A. Pharmaceutical inhalation aerosol powder dispersion—an unbalancing act. *Am Pharm Rev.* 2003;6:106–110.

24. Begat P, et al. The cohesive–adhesive balances in dry powder inhaler formulations I. Direct quantification by atomic force microscopy. *Pharm Res.* 2004;21:1591–1597.

25. Begat P, et al. The cohesive–adhesive balances in dry powder inhaler formulations II: influence on fine particle delivery characteristics. *Pharm Res.* 2004;21:1826–1833.

26. Crowder T, Donovan M. Science and technology of dry powder inhalers. In: Smyth H, Hickey A, eds. *Controlled Pulmonary Drug Delivery*. New York: Springer; 2011:203–222.

27. Hickey A, Mansour H. Formulation challenges of powders for the delivery of small molecular weight molecules as aerosols. In: Rathbone M, et al., eds. *Modified-release Drug Delivery Technology*. New York: Marcel Dekker; 2008.

28. Boer Ad, Eber E. Pulmonary. In: Bowman-Boer Y, Fenton-May V, Brun PL, eds. *Practical Pharmaceutics: International Guideline for the Preparation, Care and Use of Medicinal Products*. New York: Springer; 2015:99–130.

29. Dunbar C, et al. A comparison of dry powder inhaler dose delivery characteristics using a power criterion. *PDA J Pharm Sci Technol.* 2000;54:478–484.

30. Louey M, Oort MV, Hickey A. Standardized entrainment tubes for the evaluation of pharmaceutical dry powder dispersion. *J Pharm Sci.* 2006;37:1520–1531.

31. Coates M, et al. Effect of design on the performance of a dry powder inhaler using computational fluid dynamics. Part 1: Grid structure and mouthpiece length. *J Pharm Sci.* 2004;93:2863–2876.

32. Szefler S, Eigen H. Budesonide inhalation suspension: a nebulized corticosteroid for persistent asthma. *J Allergy and Clin Immunol.* 2002;109:730–742.

33. Niven R, Hickey A. Atomization and nebulizers. In: Hickey A, ed. *Inhalation Aerosols, Physical and Biological Basis for Therapy*, Second ed. New York: Informa Healthcare; 2007:253–283.

34. Lefebvre A. *Atomization and Sprays*. New York: Hemisphere Publishing Corporation (Taylor and Francis Group); 1989.

35. Berglund R, Liu B. Generation of monodisperse aerosol standards. *Environ Sci Technol.* 1973;7:147–153.

36. Pham S, et al. In-vitro characterization of the eFlow closed-system (eFLow CS) nebulizer with glycopyrrolate inhalation solution (SUN-101). *Am J Respir Crit Care Med.* 2017;195: A5472.

37. Sagalla R, Smaldone G. Capturing the efficiency of vibrating mesh nebulizers: minimizing upper airway deposition. *J Aerosol Med Pulmonary Drug Delivery.* 2014;27:341–348.

38. Lavorini F. The challenge of delivering therapeutic aerosols to asthma patients. *ISRN Allergy.* 2013;2013:102418.

39. Hassan A, et al. In vitro/in vivo comparison of inhaled salbutamol dose delivered by jet nebulizer, vibrating mesh nebulizer and metered dose inhaler with spacer during non-invasive ventilation. *Exp Lung Res.* 2017;43:19–28.

40. Brand P, et al. Higher lung deposition with Respimat soft mist inhaler than HFA-MDI in COPD patients with poor technique. *Int J Chron Obstruct Pulmon Dis.* 2008;3:763–770.

41. Zierenberg B, Eicher J, Dunne S. Boehringer Ingelheim nebulizer BINEB. A new approach to inhalation therapy. In: Dalby R, Byron P, Farr S, eds. *Respiratory Drug Delivery V*. Buffalo Grove, IL: Interpharm Press; 1996:187–193.

42. Newman S, et al. The BINEB (final prototype): a novel hand-held multidose nebuliser evaluated by gamma scintigraphy. *Eur Respir J.* 1996;9:441S.

43. Thurston R, et al. In: USPTO, ed. *Compositions for Aerosolization and Inhalation*. USA: Battellepharma, Inc; 2003.

44. Morad M, et al. A very stable high throughput Taylor cone-jet in electrohydrodynamics. *Sci Rep.* 2016;6:38509.

45. Sommerville M, et al. Lecithin inverse microemulsion for the pulmonary delivery of polar compounds utilizing dimethylether and propane as propellants. *Pharm Dev Tech.* 2000;5:219−230.

46. Sommerville M, Hickey A. Aerosol generation by metered-dose inhalers containing dimethylether/propane inverse microemulsions. *AAPS PharmSciTech.* 2003;4:E58.

47. Sommerville M, et al. *Lecithin microemulsions in dimethyl ether and propane for the generation of pharmaceutical aerosols containing polar solutes. Pharm Dev Tech.* 2002;7:273−288.

48. Cipolla D, et al. Lipid based carriers for pulmonary products: preclinical development and case studies in humans. *Adv Drug Delivery Rev.* 2014;75:53−80.

49. Geller D, Weers J, Heuerding S. Development of an inhaled dry-powder formulation of tobramycin using pulmosphere technology. *J Aerosol Med Pulmonary Drug Delivery.* 2011;24:175−182.

50. Cipolla D, Blanchard J, Gonda I. Development of liposomal ciprofloxacin to treat lung infections. *Pharmaceutics.* 2016;8:6.

51. Olivier K, et al. Inhaled amikacin for treatment of refractory pulmonary nontuberculous mycobacterial disease. *Ann Am Thorac Soc.* 2014;11:30−35.

52. Sirand C, Varlet J-P, Hickey A. Aerosol-filling equipment for the preparation of pressurized-pack pharmaceutical formulations. In: Hickey A, ed. *Pharmaceutical Inhalation Aerosol Technology.* New York: Marcel Dekker; 2004:331−343.

53. Hickey A, Misra A, Fourie P. Dry powder antibiotic aerosol product development: Inhaled therapy for tuberculosis. *J Pharm Sci.* 2013;102:3900−3907.

54. Pharmacopeia, *U.S., General Chapter <601> Aerosols, nasal sprays, metered dose inhalers, and dry powder inhalers.* 2011, USP Rockville, MD: United States Pharmacopeia. 218−239.

55. Sbirlea-Apiou G, et al. Bioequivalence of inhaled drugs: fundamentals, challenges and perspectives. *Ther Delivery.* 2013;4:343−367.

CHAPTER 3

Quality and Performance Tests

The overall performance of orally inhaled drug products (OIDPs) is a function of the formulation, device components, their interaction, and the capacity to accurately and reproducibly generate a pharmaceutical aerosol suitable for pulmonary delivery.[1]

The chemical composition of the active and inactive components of the formulation and their interaction create a stable milieu that can be dispensed and dispersed from the device into the aerosol. The formulation may be a solution or suspension in aqueous or nonaqueous medium or a dry powder. The device needs to actively impart energy for aerosol generation [pressurized metered dose inhalers (pMDIs), nebulizers, soft mist inhalers] or respond to energy imparted by the inspiratory flow of the patient [dry powder inhalers (DPIs)] to achieve the same objective.[2]

Unlike most aerosols with which we have experience—ambient, environmental, occupational—pharmaceutical aerosols are transient nonequilibrium phenomena that only come into existence when the inhaler is actuated and have a limited existence in the period following generation. The coordination with the patient's inspiratory flow is of the greatest importance to achieving the desired therapeutic effect. Assuming this coordination is possible, the aerodynamic particle size distribution (APSD) and the mass of the drug delivered (dose) in the respirable size range (fine particle mass) are both measures of the quality of the product and in principle are links to the prospects of achieving the desired therapeutic outcome.

Fig. 3.1 illustrates the general considerations that must be addressed for the chemistry, manufacturing, and controls (CMC) portion of a regulatory submission.

Inhaled Pharmaceutical Product Development Perspectives. DOI: https://doi.org/10.1016/B978-0-12-812209-9.00003-8

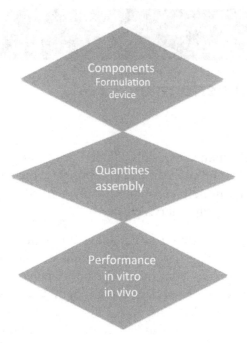

Figure 3.1 Key CMC properties, quality, and performance considerations for inhaled product development.

PHYSICAL TESTS FOR DRUG AND EXCIPIENT CHARACTERIZATION

Drugs and excipients are manufactured and processed prior to use, and their full characterization is required to establish specifications on quality that allow sufficient control of the product to guarantee its performance. Performance relates to the safety and efficacy of the product, and is based on the quality control and assurance established during development.[3] In the current regulatory environment, there is an explicit requirement that manufacturers will adopt quality-by-design principles to support the product's performance.[4]

The quality of the drug substance must be established in terms of chemical composition and physical form. For drug in the solid state, various methods of characterization are required to establish different properties that are important to the performance of the product. Fig. 3.2 shows the physical tests that are employed to characterize properties and performance of OIDPs.

The morphology of the drug can be established by microscopy. Since the final form of the drug is in the micron size, optical

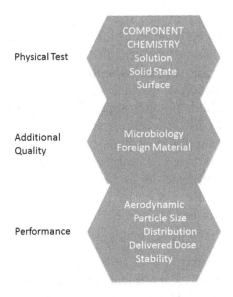

Figure 3.2 Physical tests used to characterize properties and performance of OIDPs.

microscopy has insufficient magnification. Consequently, scanning electron microscopy is the preferred method for imaging particles. In addition, it can be used as a complementary method to indirect measures of particle size and distribution. It should be noted that microscopy is limited by the arduous nature of measuring the size of individual particles and the frequently limited numbers of particles in any field of view. While image analysis tools improve estimates, the final measures are most useful in conjunction with alternative population-based methods such as laser diffraction. Also, sizing by microscopy yields populations in terms of numbers of particles, which is less useful than expressing the distribution in terms of volume or mass. For log-normal distributions, conversion from number distributions to volume or mass can be performed using Hatch–Choate equations.[5]

The composition or homogeneity of powder mixtures can be studied by X-ray microanalysis, which is often a feature of scanning electron microscopes. Marker atoms associated with the drug can be monitored to indicate their presence in a blend or for surface impurities, where the marker atom would be absent.

Laser diffraction particle size analysis to characterize drug substance alone can be done in a nonsolvent suspending agent or by high-pressure dispersion of the drug in air.[6,7] Laser diffraction uses the principle of degree of surface curvature in differentially diffracting laser light as a function of particle size. Individual particles are not imaged by this method. Instead, populations of particles diffract light onto detectors set at different radii from the axis of the laser in a manner that through calibration defines the light falling on the detector with respect to the particle size range it represents. By plotting the light falling on a series of detectors arranged in this manner, an accurate particle size distribution can be reconstructed. Both number and volume distributions can be obtained by this method, but it is more common to express the data as a volume distribution.

The solid-state chemistry that underlies particle appearance may be considered in terms of its molecular structure. In many cases, this is ordered and crystalline.[8] However, structure may be influenced by processing to introduce a less ordered, amorphous content. The bulk properties may be explored by X-ray powder diffraction.[9] The reflection of X-rays impinging on the surface of a crystal has been shown to correlate with the lattice spacing and allows reconstruction of the three-dimensional matrix of molecules of which the particle is composed. A powder diffractogram consists of a series of peaks, each corresponding to a structural feature; the amplitude and frequency of these peaks relates to the degree of crystallinity, and the absence of such features suggests a lack of structure and the presence of amorphous material.

The structural heterogeneity of some molecules leads to the potential for more than one crystal system to occur, in which case the various forms are polymorphs. Since the lattice spacing underpinning the particle matrix varies for each polymorph, the energy binding the molecules together differs. The melting point, that point at which the solid becomes a liquid, differs for polymorphs. Consequently, thermal analysis can be used to determine the existence of polymorphs. Differential scanning calorimetry raises the temperature of two containers, one containing the solid and the other empty. As the solid melts, energy is absorbed to overcome the molecular interaction in the solid. As energy is absorbed, the temperature does not increase. By comparison with the empty container, which continues to rise in temperature, a temperature differential is created which is proportional to the enthalpy required to move between states. The temperature at which this change occurs differs from one polymorph to another.

Water can take several forms in particulate materials. It may be part of the crystal lattice; it may be bound in some other, nonspecific manner; or it may be freely associated, usually with the surface of the particle. The presence of water may give rise to interaction between particles by capillary forces where the interfacial tension of water, which is high due to hydrogen bonding, gives rise to a so-called suction potential. The total water content can be determined by Karl Fischer titrimetric analysis. Thermogravimetric analysis measures the weight loss on heating and indicates the proportion of free and bound water, since the temperature required to drive them off a solid varies, free water being liberated at approximately 100°C and bound water up to 150°C.

The interaction between particles is of great importance to product performance. Consequently, the surface properties of the particles must be measured and where possible controlled. There are many surface analytical tools that may be applied, but none has yet been adopted as a standard method. Two that might be cited for their value as development methods are dynamic vapor sorption and atomic force microscopy. Dynamic vapor sorption is used to probe the interaction of the particles with moisture at a range of relative humidities and has implication for the stability of the powder in storage with respect to particle interactions (capillary forces) and other changes in particle properties (recrystallization of amorphous regions). Atomic force microscopy has been used to probe particle interactions by looking at detachment forces and variations that exist particularly in dry powder blends.

Other methods are inverse-gas chromatography,[10] Raman spectroscopy,[11] secondary ion mass spectroscopy, and X-ray photoelectron spectroscopy.[12] All of these explore the composition of the solid particles in the formulation.

Nebulizer solutions are subject to the same considerations as parenteral solutions with regard to pH, ionic strength, and their impact on solution stability.[13] Other properties, such as surface tension and viscosity, affect the performance of nebulizer droplet formation.[14] Nebulizer suspensions, such as the steroid budesonide, are prepared as colloidal suspensions, which are known to be stable due to the finely divided state of the suspended nanoparticles. The exceedingly small particle size does not influence the droplet production or the APSD. As the suspended particle size distribution approaches the droplet size (median 1–3 μm for air jet systems), it may influence the final droplet size delivered, which detracts from the use of micron sized suspension formulations.

The nonaqueous solutions and suspensions delivered from pMDIs require consideration of the affects of cosolvent (e.g., ethanol), surfactant (e.g., oleic acid), moisture, and extractables and leachables from components of the packaging (e.g., elastomers and plastics).

AERODYNAMIC PARTICLE SIZE DISTRIBUTION

The APSD of pharmaceutical aerosols describes the mass of drug in a range of particle sizes produced by the inhaler. It is important in two respects. First, aerodynamic diameter has been correlated with lung deposition through a variety of mechanisms. The aerodynamic diameter is defined as the diameter of a unit density sphere with the same terminal settling velocity as the observed particle. It differs from the geometric diameter according to Stokes' Law, which equates particle behavior in terms of terminal settling velocity (V) and accounts for both shape and density as follows:[15]

$$\frac{D_a^2 \, \rho_0 \, g \, (C_a)}{18 \, \eta} = \frac{D_g^2 \, \rho_P \, g \, (C_g)}{18 \, \eta \, \chi} = V \qquad (3.1)$$

The terms in this expression are aerodynamic and geometric diameters (D_a and D_g), unit (1 g/mL) and true particle densities (ρ_0 and ρ_p), aerodynamic and geometric slip correction factors (C_a and C_g), shape factor (χ), and gas (air) viscosity (η).

This expression can be simplified for near-spherical particles, >1 μm in diameter, to the following:

$$D_a = Dg \, \sqrt{\rho_P} \qquad (3.2)$$

This description of individual particle diameters must then be considered in the context of populations of particles, and there are several physical attributes of the particle populations that can be used to express the distribution. Clearly, the number of particles of a specific size may be used to describe the distribution. This is usually the approach to measuring distributions of sizes by microscopy. Alternatively, the volume of particles of a specific size may be considered, as is the case with optical methods such as laser diffraction particle size measurement. However, mass relates directly to dose of drug. The mass of particles of a specific aerodynamic size may be combined with the mass of all other sizes in the range of sizes and used to depict the total distribution of mass as a function of size. This APSD is the most relevant to pharmaceutical aerosols. Inertial (cascade) impaction is the only method that allows sampling of the entire aerosol according to aerodynamic particle size with a chemical detection method to measure mass of drug deposited at each stage. The sequence of decreasing orifice sizes at each of a series of stages within the sampler leads to the common term for the method, cascade impaction.[16, 17] Fig. 3.3 illustrates the principle of inertial sampling and cascade impaction. It is important to note that this method is not based on a direct measurement technique such as visual imaging (microscopy) or on fundamental properties of the particle such as light scattering but rather on semiempirical expressions for particle behavior in moving air. Consequently, cascade impactors are samplers whose accuracy and precision in estimating particle size are only as good as their calibration, which requires lengthy experimental measurements.

Stokes' number as used to define the collection efficiency of stage [18] is derived by

$$Stk = \frac{\tau U}{Dj/2} = \frac{\rho p d 2 p U Cc}{9 \, \eta Dj} \tag{3.3}$$

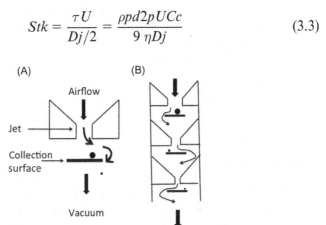

Figure 3.3 (A) General principle of inertial sampling and (B) sequential collection principle of cascade impaction.

where τ is relaxation time, ρp is particle density, η is the viscosity of the gas (air), and U is the linear velocity (volumetric airflow rate divided by the cross-sectional area diameter of the jet). For circular jets, the collection efficiency curve has a d50 cut-off at a value of ~ 0.5 for $\sqrt{\text{Stk}}$.

Calibration experiments require that each stage be studied with respect to its ability to collect a range of monodisperse aerosols from which its collection efficiency can be determined. In principle, there is a size above which approximately 100% of the aerosol is collected and below which none of the aerosol is collected. That is, at a larger size everything is collected on the stage, and at smaller sizes particles can pass around the collection surface due to insufficient inertia to proceed in the initial direction of flow.

Calibration of cascade impactors is considered so time-consuming that the convention has become to establish specifications on key dimensions of a calibrated impactor (orifice diameter, jet to collection surface distance) known to correlate with a defined cut-off diameter.[19] From these specifications, stages can be mensurated—that is, their physical dimensions can be controlled—from which it is inferred that they would give similar calibrated cut-off diameters were they to be subjected to a formal calibration study.

Cascade impactors are calibrated at fixed flow rates controlled by a vacuum pump situated at the outlet of the device. Since the operation of the instrument is dependent on the movement of air, its set-up also plays an important role in performance. Seals between each stage must be checked to avoid leaks, which would affect the calibration. A solenoid is usually employed to control the operation of the vacuum pump and the period of sampling. It may be important to ground the impactor, particularly for dry powders, where an electrostatic charge could influence behavior.

DELIVERED DOSE AND UNIFORMITY

The dose delivered from an inhaler is the proportion of the nominal dose that leaves the mouthpiece and is sometimes referred to as the emitted dose. While all drug delivered from an inhaler is captured in APSD determinations, the confounding sampling and analytical errors that occur by dividing the dose into recovered size fractions and summing measurements is undesirable. Consequently, a separate method for sampling the entire delivered dose in a single measurement is analytically preferred and from a regulatory standpoint is required.

Methods for measuring delivered dose and uniformity must be validated, and several approaches have been adopted. Three methods have seen application, but not all have appeared in pharmacopeia or regulatory guidances. An early system involved the actuation of the aerosol through a tube and sintered glass immersed in the medium in which the drug is dissolve, the aerosol traveling on airflow generated from a vacuum pump (Fig. 3.4A).[20] This compendial method was replaced by a Teflon tube through which the aerosol was delivered onto a filter on vacuum airflow (Fig. 3.4B).[21] Other methods have been proposed for either analytical or practical reasons, such as the use of separatory funnels.

As DPIs became more common, the impact of their operational features on performance was of increasing interest. Unlike pMDIs, where the drug is administered by the propulsive force of the evaporating propellant, DPIs use the inspiratory flow of the patient to disperse the drug. Early inhalers such as the Rotahaler and Spinhaler offered little resistance, but as more inhalers came to the market it became clear that resistance varied significantly between different manufacturers and that flow through the device varied as a consequence. This observation led to the requirement that all delivered dose measurements should be conducted at pressure drop of 4 kPa across the device. Additionally, it was noted that to sample from the inhaler for an arbitrarily long period (sample volume) lacked relevance to the use of the device. While the context of this observation was not clinical, it was evident

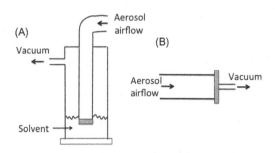

Figure 3.4 (A) Sintered glass impinger and (B) dose uniformity sampling apparatus.

that a quality measure of performance periods of sampling and volumes of air sampled should be considered important variables.

OTHER FACTORS

To improve the prospects of fully understanding and characterizing new drug products and facilitating generic drug product development, other areas in which measurement might be made to establish the quality of the product have been under investigation.

The response of inhaler performance to airflow has been acknowledged in the research on delivery of drugs at different flow rates and different pressure drops. Many researchers have studied the effect of flow conditions[22] such as those specified for APSD testing in the US Food and Drug Administration (FDA) guidance document on metered dosing of 30, 60, and 90 L/min,[23] or of pressure drop to study the formulation behavior using laser diffraction,[24] or by different devices.[22] All approaches are independent of the patient's inspiratory flow profile.

An alternative is to retain the ability to sample at the calibrated flow rate of the impactor but change the pressure drop in a controlled manner at the point of aerosol delivery. This can be achieved by adopting different inhalers with a range of pressure drops or by using a more uniform approach with the standardized entrainment tubes (SETs).[25] Fig. 3.5 shows the way in which SETs can be constructed

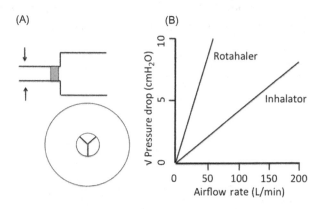

Figure 3.5 (A) Standardized entrainment tube (SET) and (B) pressure drop as a function of airflow for a variety of inhalers and SETs.

with a range of pressure drops (Reynolds' numbers, shear stresses) to cover the performance of marketed inhalers. The diameter of the tube can be varied as indicated by the arrows and a series of tubes with different flow properties constructed. The SET data are relevant and comparable since additional features such as grids, tortuous channels, and impaction surfaces designed into inhalers for a specific formulation are not present in the SETs. The data may be useful as a guide to selecting the required conditions for dispersion of a formulation to design a device and reduce the amount of parallel development required.

It has been common practice to control either the airflow rate or the pressure drop in evaluating the performance of DPIs. This is understandable because these are measures of the inspiratory capacity of the patient (airflow) and the response of the device to the effort expended (pressure drop). However, these two measures are clearly linked in practice, because the pressure drop is influenced by and in turn influences the inspiratory airflow. The link between these two measures can be recognized by the adoption of another familiar term, power. The power required to disperse the powder can be calculated from the pressure drop (ΔP) and the airflow (Q) according to the following expression:[26]

$$Power = \Delta P \cdot Q \qquad (3.4)$$

Adopting this, power performance criterion more accurately reflects the entire phenomenon involved in the dispersion of powders. As airflow profiles that do not exhibit constant conditions are adopted, this may become a valuable reference term.

IVIVC CONSIDERATIONS

It is not surprising that a perennial interest has been expressed in cascade impaction data as an indicator of lung deposition. Since the aerodynamic diameter has been used as the primary determinant of particle deposition, models developed in the last century for occupational and environmental exposure used it. The development sequence has historically involved thorough in vitro evaluation, as described above, concluding with APSD and delivered dose estimates. This could then be used to support lung deposition imaging and pharmacokinetic and pharmacodynamic studies in man.[27] However, the in vitro indicators of appropriate performance have not consistently correlated with phar-macokinetics and efficacy to allow prediction of in vivo performance or be the foundation for bioequivalence. Occasionally, correlations have been found between literature values for the range of DPI and pMDI lung deposition as derived from gamma scintigraphy imaging and a particular measure of aerodynamic particle size.

Fig. 3.6A summarizes the findings of Newman and Chan with respect to an almost perfect correlation of lung deposition with parti-cles of aerodynamic diameter less than 3 μm by cascade impaction.[28] The dotted lines show the range in the data. However, similar correla-tions did not occur at higher and lower particle sizes.

A more complex and compelling study by Olsson et al. evaluated the relationship of measured lung deposition with respect to the pro-portion of the dose that penetrated through three different upper respi-ratory casts using different inspiratory flow profiles.[29] Fig. 3.6B shows an exact correlation of lung deposition and ex-cast deposition.

Both examples suggest that a correlation may be on the horizon. These practical endeavors are supported by the theoretical observation that lung deposition can be correlated with impaction factor.[30] Regulatory science is also adapting to the possibility of stronger links between each of the product development activities.[31]

Because lung deposition cannot be presumed to correspond with bioavailability or pharmacological action, caution should be

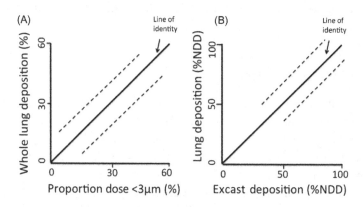

Figure 3.6 Correlations between in vivo lung deposition and (A) proportion of particles with diameters below 3 μm determined by cascade impaction[28] and (B) exoropharyngeal cavity cast deposition.[29] Dotted lines represent the range of individual data with respect to the line of unity.

exercised in anticipating an in vitro−in vivo correlation with therapeutic effect. Recently, the potential for an inhaled biopharmaceutical classification system (iBCS) was evaluated.[32] The approach is broadly based on that developed for oral products, which can be reduced to considerations of high and low solubility and permeability of the drug.[33] It is not clear that such a simple approach would be suitable for inhaled products, but if a correlation is found between APSD and lung deposition the next logical step would be an iBCS. The major consideration appears to be the kinetics of disposition in the lungs and the fact that most drugs are locally acting such that permeability may not be the most relevant measure of bioavailability. Moreover, it may be valuable to consider a variation on Lipinski's rule of 5,[34] given the different medium, with variable hydrophobicity and hydrophilicity, into which the drug is delivered in specific regions of the lungs and the residence time at that location based on dominant clearance mechanism at that site.[35]

STATISTICAL PARAMETERS

The application of statistics to the interpretation of inhaler performance data has been a subject of discussion for many years. The complexity of the problem lies in the meaning of the data being analyzed. APSD data from inertial impaction sampling is basically a series of mass determinations of recovered drug from stages or components on which aerosol is deposited. These samples arbitrarily divide the APSD, but their interpretation is intended to reflect the actual APSD, which can be described by a mean, median, and mode around which particles in a range of sizes are distributed. Consequently, the APSD is itself subject to a statistical interpretation independently of the statistics that might be applied to masses deposited on each of the stages.

If the intent is to compare two distributions for similarity, it is essential that we not only consider the deviation of masses deposited on each of the stages but the implications of these deviations for the entire estimate APSD as it reflects the actual APSD. Several groups have investigated this phenomenon.

The use of statistical methods to evaluate inertial impactor data is the subject of some debate. The mass of drug depositing on each stage of the impactor through stage calibration is used to reconstruct the

APSD. There are limitations to this method based on the dosage form, the nature of the aerosol, and the method of sampling.

Each dosage form delivers the aerosol by a unique mechanism that interacts with the sampling technique. pMDIs actively disperse drugs on the rapidly expanding plume arising from the evaporation of propellant; dry powders are dispersed on the inspiratory flow of the patient; nebulized droplets are generated as steady state aerosols on a compressed carrier gas (vibrating mesh system). As a consequence, the pMDI aerosol is rapidly moving on the constant flow drawn through the impactor. A proportion of particles, regardless of size, will likely deposit in the impactor inlet. Dry powder aerosols are drawn into the impactor on the constant flow, and particles conform to size segregation as intended for the instrument. Nebulizer output is propelled on an independent airflow that is sampled by the impactor at intervals during delivery. Consequently, as shown in Table 3.1, each dosage form, dispersion method, and sampling approach has flaws with respect to the intent of inertial impaction.

The nature of the delivery phenomena having been established, attention can be focused on the sampling conditions. Two important variables are the inlet dimensions and the airflow rate. In the late 1980s, convergence occurred on an inlet geometry that could be used as a standard for pharmaceutical aerosol characterization. The appearance of a standard inlet in the United States Pharmacopoeia in 1990

Table 3.1 Criteria to Consider When Choosing Inertial Impaction as an Aerodynamic Particle-Sizing Tool			
Feature	pMDI	DPI	Nebulizer
Carrier gas—independent of patient	Yes	Yes	Yes
Carrier gas—patient inspiratory flow	No	Yes	No
Dispersion—independent of patient inspiratory flow	Yes	No	Yes
Dispersion—dependent on patient inspiratory flow	No	Yes	No
Velocity—high	Yes	No	No
Velocity—variable	No	Yes	No
Velocity—low	No	No	Yes
Sampling principle—bolus	Yes	Yes	No
Sampling principle—steady state	No	No	Yes
Inlet deposition—particle size independent	Yes	No	No
Inlet deposition—adheres to impaction principles	No	Yes	Yes

allowed a means of collecting data for comparison across products and laboratories for the first time. In recent years, alternative geometries that approach the anatomical throat have been developed. The new systems have yet to be universally adopted but are proposed to allow sampling of physiologically relevant fractions of the aerosol into the impactor.

The flow rates employed in cascade impactors were originally intended to sample at resting or sedimentary conditions for ambient sampling of hazardous aerosols.

The standard impactor used to be the Andersen 1 ACFM (actual cubic feet per min) eight-stage nonviable inertial sampler. The airflow rate of 1 ACFM is equivalent to 28.3 L/min. Subsequently, impactors were calibrated at 30 L/min. For many years, these impactors were operated at these flow rates particularly for pMDIs and nebulizers. The observation that dry powders require dispersion dictated by flow rates led to calibration of impactors at 60 and 90 L/min to allow for higher shear conditions and the potential to observe powder response to flow. Since powders often contain carrier particles that are too large to enter the sizing stages of the impactor, a preseparator is interposed after the sampling inlet to collect the large carrier particles and aggregates.

Once the particles pass into the impactor, curvilinear airflow occurs and impaction based on Stokes' law. The principle of operation is collection of aerosol below orifices on sampling surfaces. The sequence of stages that make up the cascade has decreasing orifice sizes that, given the constant volumetric flow, means increasing linear velocity, which allows ever decreasing particle sizes to be collected. Since all variation in the dosage form, dispersion mechanism, and sampling occurs before the sizing stages, these factors should not influence the particle size estimates of the impactor sized mass.

Originally, it was assumed that APSDs were log-normally distributed, but alternate mathematical distributions are often a better fit to the data from different pharmaceutical aerosol dosage forms.[36] This observation has resulted in a greater focus on the masses deposited on each stage of the impactor rather than on mathematical fits to the data.

The Product Quality Research Institute (PQRI) evaluated statistical methods of APSDs obtained from inertial impactions data. The intent was to compare distributions and establish, with the use of statistical analyses, whether it was reasonable to assume that, within the limits of experimental variation, they could be said to be identical.[37] It was concluded that if the entire emitted dose was evaluated, there was no basis, in comparison with expert opinion, from which statistics alone could be used to demonstrate similarity. However, the chi-squared analysis applied to only the impactor sized mass did show sufficient similarity to be considered a useful tool.

A principle components analysis approach was also proposed in which the example profiles used in the PQRI evaluation were subjected to analysis with reasonable success in predicting the similarity of profiles as defined by expert opinion.[38]

The International Pharmaceutical Aerosol Consortium on Regulation and Science pursued this topic further and produced an excellent text describing the considerations related to good cascade impactor practices.[17]

STABILITY

The conditions under which stability studies are conducted have been published by the International Council for Harmonisation of Technical Requirements for Pharmaceuticals for Human Use and the US FDA.[39] Approaches to statistical design and sampling are well documented.[40] Each dosage form has its requirements for storage. The general storage conditions of temperature and relative humidity (RH) for conducting stability studies are 25°C/70%RH, 40°C/75%RH, and for powders the intermediate of 30°C/60%RH. The absence of change under these storage conditions is considered a measure of stability. Shipping and transportation studies are also required to establish that no changes in the product are likely to occur during commercial distribution.

Pressurized Metered Dose Inhalers

Due to the significance of the contact of the metering valve with the reservoir of the pMDI, the orientation of storage of the device (valve up/valve down) must be included in the conditions of storage. As the pMDI is a reservoir device, performance must be assessed through the life of the product, which can be as many as 200 doses.

Dry Powder Inhalers

DPIs take the form of unit dose, multiunit dose, and reservoir systems. The design of stability studies must assess the dosage form with respect to the number of doses the system is intended to deliver. Consequently, unit dose devices establish the storage stability of the replicates of the unit (capsule, blister). Multiunit dose devices and reservoir devices must address stability through the lifespan of the product.

Nebules

Nebulizers are not regulated as drug products, because the device and formulation are sold separately. Consequently, stability considerations are focused at the level of the dispensing unit, the nebule. The focus of stability studies is on the chemical stability of the drug and the physical stability of a suspension unless there are reasons to anticipate changes in solution or suspension properties, particularly viscosity and surface tension.

CONCLUSION

The quality and performance tests performed on OIDPs are done primarily to ensure the accuracy and reproducibility of drug delivery over the lifespan of the product. During development of the product, these properties are controlled with respect to safety and efficacy.

It is convenient to consider the properties of the product in terms of the formulation composition and device components, the quantities of specific materials in the formulation, their individual and combined physicochemical properties, interaction with metering system and device components, and the implications of all of these variables for in vitro and in vivo performance measurements.

Efforts are underway to render in vitro method predictive of in vivo effect, and some success has occurred in predicting lung deposition. Extrapolation to an iBCS is under consideration, but this may be a more difficult task.

Statistical approaches to evaluating performance measures, particularly APSD, have received considerable attention, and the ability to demonstrate profile equivalence is now being addressed by the industry and regulators.

All quality and performance tests find their ultimate utility in demonstrating the stability and robustness of the product in a variety of storage and usage conditions and its ability to support accurate and reproducible drug delivery throughout its shelf life.

REFERENCES

1. Hickey A. Summary of common approach to pharmaceutical aerosol administration. In: Hickey A, ed. *Pharmaceutical Inhalation Aerosol Technology*. Second ed. New York, NY: Informa Healthcare; 2004:385–421.

2. Clark A. Pulmonary delivery technology: recent advances and potential for the new millennium. In: Hickey A, ed. *Pharmaceutical Inhalation Aerosol Technology*. Second ed. New York, NY: Marcel Dekker; 2004:571–591.

3. Hickey A, Ganderton D. *Quality by design. Pharmaceutical Process Engineering*. Second ed. 2010:193–196.

4. Food US, Administration Drug. *Guidance for Industry: Q8(R2) Pharmaceutical Development*. Washington DC: US Department of Health and Human Services; 2009.

5. Stockham J, Fochtman E. *Particle Size Analysis*. Ann Arbor, MI: AnnArbor Science; 1977.

6. Hickey A, et al. Physical characterization of component particles included in dry powder inhalers. II Dynamic characteristics. *J Pharm Sci*. 2007;96:1302–1319.

7. Hickey A, et al. Physical characterization of component particles included in dry powder inhalers. I. Strategy review and static characteristics. *J Pharm Sci*. 2007;96:1282–1301.

8. Mullin J. *Crystallization*. Third ed. Oxford, UK: Butterworth-Heinemann; 1993.

9. Crowder T, et al. *Introduction to Pharmaceutical Particulate Science and Technology*. Boca Raton, FL: CRC Press; 2001.

10. Telko M, Hickey A. Critical assessment of inverse gas chromatography as means of assessing surface free energy and acid–base interaction of pharmaceutical powders. *J Pharm Sci*. 2007;2647–2654.

11. Mansour H, Hickey A. Raman characterization and chemical imaging of biocolloidal self-assemblies, drug delivery systems, and pulmonary inhalation aerosols: a review. *AAPS PharmSciTech*. 2010;8:E99.

12. Hickey A, Jackson G, Fildes F. Preparation and characterization of disodium fluorescein powders in association with lauric and capric acids. *J Pharm Sci*. 1988;77:804–809.

13. Carstensen J. *Solution kinetics. Drug Stability*. New York, NY: Marcel Dekker; 1990:15–108.

14. Niven R, Hickey A. Atomization and nebulizers. In: Hickey A, ed. *Inhalation Aerosols Physical and Biological Basis for Therapy*. Second ed. New York, NY: Informa Healthcare; 2007:253–283.

15. Crowder T, et al. Fundamental effects of particle morphology on lung delivery: predictions of Stokes' law and the particular relevance to dry powder inhaler formulation and development. *Pharm Res*. 2002;25:239–245.

16. Lodge J, Chan T. *Cascade Impactor*. Akron, OH: American Industrial Hygiene Association; 1986.

17. Tougas T, Mitchell J, Lyapustina S. *Good Cascade Impactor Practices, AIM and EDA for Orally Inhaled Products*. New York, NY: Springer; 2013.

18. Hinds W. *Aerosol Technology, Properties, Behavior and Measurement of Airborne Particles*. Second ed. New York, NY: John Wiley and Sons; 1999.

19. Vaughan N. The Andersen impactor: calibration, wall losses and numerical simulation. *Journal of aerosol Science*. 1989;20:67−90.

20. British Pharmaceutical Codex. 1973: Pharmaceutical Society of Great Britain, London, UK.

21. Pharmacopeia, U.S., General Chapter <601> Aerosols, Nasal Sprays, Metered Dose Inhalers, and Dry Powder Inhalers. 2011, USP Rockville, MD: United States Pharmacopeia. 218−239.

22. Coates M, et al. Influence of airflow on the performance of a dry powder inhaler using computational and experimental analyses. *Pharm Res*. 2005;22:1445−1453.

23. US Food and Drug Administration, *Draft Guidance for the industry, Metered Dose Inhaler (MDI) and Dry Powder Inhaler (DPI) Chemistry Manufacturing and Controls Documentation*. 1998.

24. Behara S, et al. Insight into pressure drop dependent efficiencies of dry powder inhalers. *Eur J Pharm*. 2012;46:142−148.

25. Louey M, Oort MV, Hickey A. Standardized entrainment tubes for the evaluation of pharmaceutical dry powder dispersion. *J Pharm Sci*. 2006;37:1520−1531.

26. Dunbar C, et al. A comparison of dry powder inhaler dose delivery characteristics using a power criterion. *PDA J Pharm Sci Technol*. 2000;54:478−484.

27. Sbirlea-Apiou G, et al. Bioequivalence of inhaled drugs: fundamentals, challenges and perspectives. *Ther Delivery*. 2013;4:343−367.

28. Newman S, Chan H-K. In vitro/In vivo comparisons in pulmonary drug delivery. *J Aerosol Med Pulmonary Drug Delivery*. 2008;21:77−84.

29. Olsson B, et al. Validation of a general in vitro approach for prediction of total lung deposition in healthy adults for pharmaceutical inhalation products. *J Aerosol Med Pulmonary Drug Delivery*. 2013;26:355−369.

30. Clark A. Understanding penetration measurements and regional lung targeting. *J Aerosol Med Pulmonary Drug Delivery*. 2012;25:179−187.

31. Forbes B, et al. In vitro testing for orally inhaled products: developments in science-based regulatory approaches. *AAPS J*. 2015;17:837−852.

32. Hastedt J, et al. Scope and relevance of a pulmonary biopharmaceutical classification system AAPS/FDA/USP Workshop March 16−17th, 2015 in Baltimore, MD. *AAPS Open*. 2016;2:1.

33. Amidon G, et al. A theoretical basis for a biopharmaceutic drug classification: the correlation of in vitro drug product dissolution and in vivo bioavailability. *Pharm Res*. 1995;12:413−420.

34. Lipinski C, et al. Experimental and computational approaches to estimate solubility and permeability in drug discovery and development settings. *Adv Drug Delivery Rev*. 2001;46:3−26.

35. Mortensen N, Hickey A. Targeting inhaled therapy beyond the lungs. *Respiration*. 2014;88:353−364.

36. Dunbar C, Hickey A. Evaluation of probability density functions to approximate particle size distribution of representative pharmaceutical aerosols. *J Aerosol Sci*. 2000;31:813−831.

37. Christopher D, et al. Product Quality Research Institute evaluation of cascade impactor profiles of pharmaceutical aerosols, Part 3: Final report on a statistical procedure for determining equivalence. *AAPS J*. 2007;8:65.

38. Shi S, Hickey A. Multivariate data analysis as a semi-quantitative tool for interpretive evaluation of comparability or equivalence of aerodynamic particle size distribution profiles. *AAPS PharmSciTech*. 2009;10:1113−1120.

39. Jones L, et al. Stability concerns and approaches to analysis of pharmaceutical aerosols. *Pharm Technol*. 2000;24:40−54.

40. Bolton S. *Quality control. Pharmaceutical Statistics*. New York, NY: Marcel Dekker; 1990:405−443.

CHAPTER 4

Preclinical and Clinical Considerations

In the previous chapters, inhaled drug products, their physicochemical properties, and metrics that define quality and performance were identified. It has long been the desire of researchers in the field to link the properties that are known to influence lung deposition, disposition, and therapeutic effect, as was briefly alluded to in Chapter 3. Fig. 4.1A illustrates the basic principle that underpins considerations on this topic. It should be possible to link the aerosol characteristics that interact with the lung anatomy and physiology to the ability to target specific receptors or other pharmacological targets involved in modulating lung function. Fig. 4.1B shows the most obvious factors in each of the categories of aerosol properties, lung logistics, and efficacy, including aerodynamic particle size distribution (APSD) and delivered dose, lung deposition and clearance, and regional deposition and disease. To the extent that even broad predictions can be made through this sequence of disposition, drug product development can be guided from established quality to the desired therapeutic outcome. However, direct correlations leading to quantitative predictions have been elusive.[1,2]

Pharmaceutical aerosol products are unique among dosage forms in that the most important elements in their delivery and performance, the aerosol particles or droplets, do not exist independently of the patient's actuation or inhalation, which is usually coordinated with the inspiratory flow cycle to maximize deposition in the lungs.[3-5] The role of the interface of the device with the patient has enormous significance for efficacy. The performance of the formulation and device in producing an aerosol has already been discussed.

The ability of the patient to expend sufficient effort to generate an aerosol from a dry powder inhaler (DPI), in which the aerosol is generated on the inspiratory flow, is of particular concern in effective delivery. Historically, there have been two independent parameters that have been the focus of dry powder dispersion considerations, flow rate

Inhaled Pharmaceutical Product Development Perspectives. DOI: https://doi.org/10.1016/B978-0-12-812209-9.00004-X

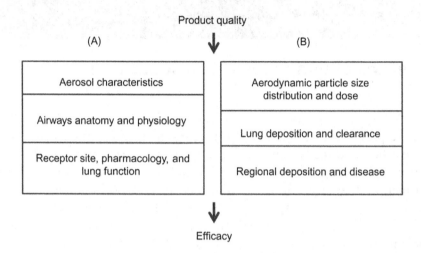

Figure 4.1 *Physicochemical and biological sequence in translating quality (aerosol performance measures) into efficacy (therapeutic outcomes): (A) general and (B) specific considerations.*

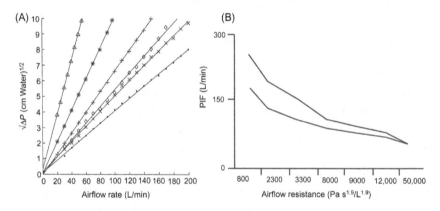

Figure 4.2 *(A) Square root of pressure drop ([cm $H_2O]^{1/2}$) versus air flow rate (L/min) for six dry powder inhaler devices (• Rotahaler, X Spinhaler, ◇ ISF Inhaler, + Diskhaler, * Turbuhaler, △ Inhalator); (B) the range for seven human volunteers of peak inspiratory air flow (PIF) rate (L/min) versus airflow resistance.* Modified from Clark, A. and A. Hollingworth, *The relationship between powder inhaler resistance and peak inspiratory conditions in healthy volunteers—implications for in vitro testing.* J Aerosol Med, 1993. **6**: p. 99–110; Olsson, B. and L. Asking, *Critical aspects of the function of inspiratory flow driven inhalers.* J Aerosol Med, 1994. **7**(S1): p. S43–S47.

generated through the device and pressure drop created across the device. The latter reflects resistance to patient effort in generating a particular airflow. Fig. 4.2A illustrates a range of pressure drops (ΔP, expressed as $\sqrt{\Delta P}$ for linearity) at a given air flow rate (Q) that can be generated through known devices.[6] The extremes of performance for the Rotahaler and the Inhalator represent low and high resistance devices, respectively. Resistance is defined as $([\Delta P]^{1/2})/Q$. The patient would experience less difficulty inhaling through the low than through the high resistance device. This raises the question of how much difficulty would the patient experience inhaling through devices of different resistances. Fig. 4.2B shows the range of peak inspiratory flow (PIF) for seven human volunteers as a function of resistance. An almost linear drop in PIF rate occurs as resistance is increased.[7] Interestingly, between-patient variations in performance are lower at higher resistance. It is important to recognize that PIF is one measure of the inspiratory flow cycle. Fig. 4.3A illustrates how the PIF can vary in magnitude based on effort and time of attainment in the cycle, which would relate to conveyance of the aerosol generated.[6] Fig. 4.3B illustrates the impact of modulating airflow on the aerosol generated. The cumulative APSD (absolute value on the abscissa, percent undersize particles on the ordinate) is shown for three air flow rates drawn through the device, in this case the Turbuhaler.[7]

In summary, the pressure-drop generated across the device influences the airflow that can be drawn through it. In turn, this may affect the APSD that is created under the conditions a particular patient can generate. Ideally, the aerosol product performance is independent of flow rate and pressure drop. Considering these terms separately is confusing since they are dependent variables. The interconnectedness of flow rate and pressure drop can be employed to describe the effort that the patient expends in generating the aerosol. The power function is expressed as $\Delta P \cdot Q$ and can be used to normalize data into a single term of clinical relevance, since it reflects patient effort.[8]

Fig. 4.4 returns to the theme of the linking strategy between product quality and efficacy. A continuing research effort is expended to investigate the potential to use in vitro aerodynamic particle size and delivered dose data to predict in vivo effect, and there are two possible links between the theoretical and experimental lung deposition and subsequent pharmacokinetic data.[9] However, for locally acting drugs, pharmacokinetic data is an indirect measurement. There is a large body of knowledge on the impact of aerodynamic particle size on lung deposition, but almost all of it was generated from passive inhalation of steady state or near-equilibrium ambient aerosols, which differ significantly from the majority of nonequilibrium pharmaceutical

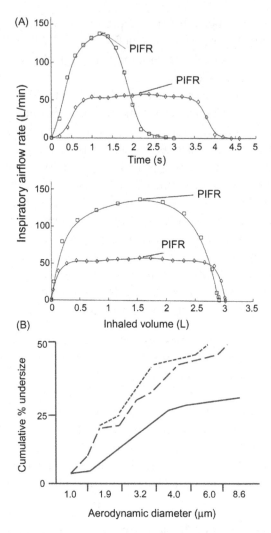

Figure 4.3 (A) Inspiratory air flow rate versus time and inhaled volume for maximum (□) and comfortable (◇) human subject effort (PIFR, peak inspiratory flow rate in each case) through the Turbuhaler; (B) APSD (% undersize) with respect to air flow rate at which sampling was conducted (continuous—40, dashed—60, dotted—80 L/min). Modified from Clark, A. and A. Hollingworth, The relationship between powder inhaler resistance and peak inspiratory conditions in healthy volunteers—implications for in vitro testing. J Aerosol Med, 1993. **6**: p. 99–110; Olsson, B. and L. Asking, Critical aspects of the function of inspiratory flow driven inhalers. J Aerosol Med, 1994. **7(S1)**: p. S43–S47.

aerosols. Moreover, the small volumes, whether considering the oropharynx or the use of spacer systems, into which inhaled aerosols are generated require consideration of velocity and acceleration in the ramp-up and down portions of the inspiratory flow cycle, as well as plume geometry and directionality. Each of these considerations

Product quality

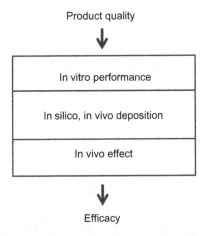

Efficacy

Figure 4.4 Physicochemical and biological sequences in translating quality into efficacy with respect to the use of in silico and in vivo deposition to bridge from in vitro testing to in vivo effect.

complicates deposition mechanisms by bringing particles or droplets into close proximity to surfaces under a range of airflow conditions that are not easily mimicked in vitro.

LUNG DEPOSITION IMAGING

It is generally acknowledged that the target particle size distribution to effectively deliver drugs to the lungs is 1 to 5 μm.[10] This number is derived from many years of data gathered in the environmental and occupational health arena, supplemented recently by clinical trials for pharmaceutical aerosols.

In considering how best to assess site and reproducibility of delivery of drugs to the lungs, there are several complementary approaches. The most prominent are two-dimensional planar gamma scintigraphy imaging, positron emission tomography, and single photon emission computed tomography (SPECT).[11,12] Fig. 4.5 shows images of left (L) and right (R) lungs produced by (A) classical gamma scintigraphy and (B) SPECT. ^{99m}Tc-DTPA is the radiolabel employed routinely. Its short half-life of 6 hours renders it useful for this purpose, as large doses are not required and the residue does not represent a serious risk to technical staff. The three gamma scintigraphy panels in Fig. 4.5A shows the following for healthy lungs: (i) an aerosol penetrated to all areas as demonstrated by the gamma scintigraphy image generated following deposition of an aerosol with an activity median diameter of

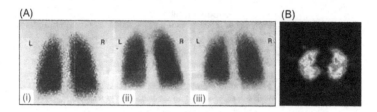

Figure 4.5 (A) Gamma scintigraphy 2D planar images for (i) ^{99m}Tc-DTPA aerosol, (ii) ^{81m}Kr gas imaging the airways, and (iii) injected ^{99m}Tc-macroaggregated albumin imaging the vasculature[13]; (B) transverse SPECT image.[14]

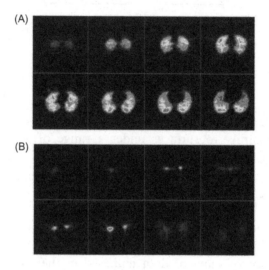

Figure 4.6 Sequential SPECT transverse sections through the lungs of a healthy (A) and an asthmatic (B) individual.[14]

0.9 μm and a geometric standard deviation of 1.5, with 96% of the aerosol below 2 μm; (ii) the lung periphery shown by independent radioactive gas imaging; and (iii) the area outlined by independent injected macroaggregated albumin-^{99m}Tc image of the vascular system.[13] The SPECT image Fig. 4.5B is a transverse section, which can be made at any plane in the lung from which a 3D image can be reconstructed.[14] The color scheme is the commonly used "hot spot" approach where red represents the most radioactivity and therefore the greatest deposition; the spectrum then ranges through yellow and green to blue, the lowest radioactivity and least deposition.

Fig. 4.6 shows eight sequential transverse lung "slices" for a healthy (A) and an asthmatic (B) patient, indicating the poor deposition in the diseased versus the healthy lungs.[14] The impact of the much lesser penetration of the aerosol into the asthmatic lungs would clearly be a lower probability of reaching areas that require treatment. However, it should be noted that as treatment progresses, the effectiveness of the therapy would result in improved penetration into the airways and increased control of both the symptoms and underlying cause of disease.[15] The observation of the synergy between sequential dosing in a regimen and progressive improvement in lung function is central to concerns over patient compliance and adherence.

There are many variables in product performance and patient biology that contribute to lung deposition and therapeutic outcome. Consequently, the potential to correlate in vitro and in vivo performance measures has to be approached with caution.

LUNG DEPOSITION MODELING

A simple approach advocated many years ago by Rudolph was to consider the entire aerosol that passes through the oropharynx to be delivered to the lungs.[16,17] If this approach is taken, then only head oropharyngeal deposition needs to be considered, and that can easily be calculated. Subsequently, it was shown that the theoretical model accurately approximated experimental data obtained for a DPI.[18]

Newman and Chan collected APSD data from several pharmaceutical aerosol dosage forms for which lung deposition data were available.[19] They then explored the possibility of a correlation between lung deposition and a particular stage-cutoff diameter of the inertial impactor used to measure the APSD. There was no correlation for most stages, but for particles <3 μm the data clustered around the line of identity, suggesting that a correlation existed as described in Chapter 3 (Fig. 3.6A).

A correlation between lung deposition and impactor sampling with respect to 3 μm particles and smaller is not surprising. Mitchell and Dunbar overlaid regional lung deposition data as a function of particle size on the Andersen eight-stage impactor calibration profiles, as shown in Fig. 4.7.[20] Clearly, extrathoracic and tracheobronchial deposition have fallen to near zero at a cutoff of approximately 3 μm. Below this point, all deposition should be peripheral and there would be no barrier to its entering the lungs.

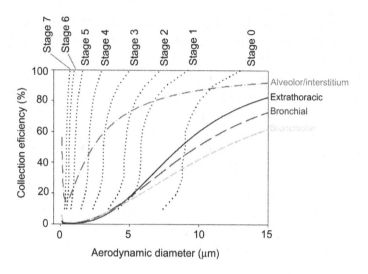

Figure 4.7 Regional lung deposition and Andersen eight-stage sampler collection efficiency curves as a function of aerodynamic particle size.[20]

It should also be noted that the impact of particle size distribution on peripheral deposition exhibits convergence at 1 and 5 μm over a wide range of theoretical geometric standard deviations (1, 2, and 3.5), representing monodispersity to extreme polydispersity.[10,21]

Since the model used to make this observation employed assumptions of passive inhalation, it would not be surprising if the particle size at which this convergence occurred for therapeutic aerosols and cascade impactors was slightly lower, such as 3 μm, because throat and sampling inlet deposition will be exaggerated due to the velocity and breathing manoeuver used to inhale the aerosol.

It is not clear that experimental correlations with regional deposition within the lungs could improve on the total lung and 3 μm apparent relationships described. However, sophisticated theoretical lung deposition models might give sufficient accuracy and reproducibility for greater resolution to predictions of regional deposition.

One computer model explored a bifurcating system at multiple levels of scrutiny to construct a whole lung (Fig. 4.8).[14,22] Using this model, deposition was compared to published deposition data (Fig. 4.9).[22,23] The figure depicts deposition in the tracheal, tracheobronchial, and pulmonary regions of a healthy male under specific conditions of pulmonary function. The predicted and observed data

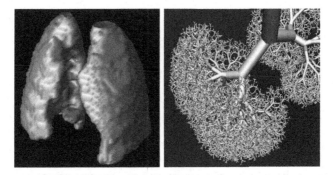

Figure 4.8 Bifurcating 3D simulated lung computer model.[14]

are very close; and as more data becomes available with which to compare and refine the theoretical model, this and related methods might be useful in product development.

While theoretical approaches have evolved, practical methods for sampling for in vitro aerodynamic particle size characterization have also been improved with respect to physiological relevance. The US Pharmacopeia sampling inlet was the first standardized one for conducting aerodynamic particle size characterization in the United States. The adoption of a standardized inlet was a major advance in that comparisons could be made from laboratory to laboratory and methods in general could be standardized. However, the USP inlet is a simple right-angled tube. The design brief did not require anatomical features of relevance to humans or performance equivalent to the human oropharynx for collection of aerosol particles and droplets. Based on earlier observations, adopting an inlet that samples the aerosol that would ordinarily deposit in the mouth and throat might be the simplest way to predict from in vitro measurements the potential for lung deposited dose. The task of considering a physiologically relevant inlet was addressed by Finlay et al.[24,25] The outcome of this research was the Alberta Throat.[26]

Others have recently explored this concept further by adopting three physiologically relevant inlets representing different ages and conducting aerosol sampling for comparison with lung deposition data. As discussed in Chapter 3, this is a promising approach to linking in vitro performance assessment with expectations of lung deposition.[27,28] A correlation was discovered for experimental lung deposition and excast drug aerosol collected after passing through the sampling inlets,

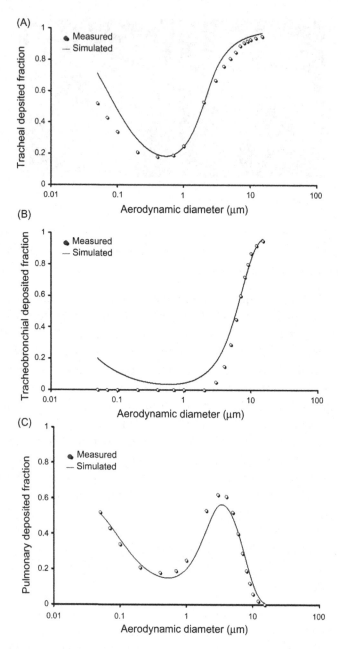

Figure 4.9 Theoretical and experimental regional lung deposition curves for (A) tracheal, (B) tracheobronchial, and (C) pulmonary regions of a healthy male. Breathing conditions: tidal volume 1 L, frequency 7.5 breaths/min, airflow rate 250 mL/s. Modified from Martonen, T., et al., Lung models: strengths and limitations. Respir Care, 2000. 45: p. 712–736; Heyder, J., et al., Deposition of particles in the human respiratory tract in the size range 0.005–15 μm. J Aerosol Sci, 1986. 17: p. 811–825.

based on oropharyngeal casts (Fig. 3.6B). Nine products, including particle metered dose inhalers (pMDIs), DPIs, and nebulizers, were assessed for their performance in anatomically correct sampling inlets. The inlets represented the delivery conditions in trained healthy individuals, with no exhaled mass. Variation in excast sampling, either for fine particle dose onto a filter or into a cascade impactor operated at a fixed flow rate with a mixing inlet to allow for variable flow through the inhaler, predicted lung deposited dose estimated independently by pharmacokinetic measurement.

While the link between in vitro measures of aerosol performance and both theoretical and experimental in vivo deposition data appear to be improving, the step to predicting in vivo effect is complicated by other biological factors. Fig. 4.10 highlights the importance of clearance mechanisms as a corollary to deposition in influencing the availability of the drug to reach the target site (pharmacodynamics) and to be transported to the systemic circulation (pharmacokinetics).[29]

The mechanisms of clearance are absorption, mucociliary transport, and cell-mediated transport. Each of these mechanisms is regionally specialized in the lungs and has evolved to respond to very specific

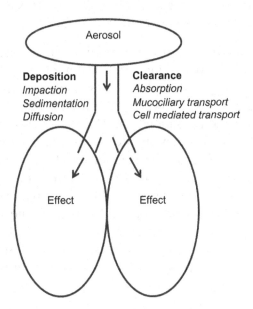

Figure 4.10 Deposition and clearance mechanisms from the lungs define the availability of drug aerosols to achieve their therapeutic effect.

forms of inhaled materials with the objective of protecting the peripheral regions in which gaseous exchange occurs.

ANIMAL MODELS

Animal models are used for many purposes in the development of inhaled drug products. Their application may be summarized as classical absorption, distribution, metabolism, elimination, and toxicology (ADMET) studies to demonstrate pharmacokinetics and safety, and pharmacodynamic or disease models to demonstrate efficacy.[30] For safety evaluation of inhaled products, the convention is to use a rodent species, usually the rat, and a larger species, usually the dog.[31,32] Animal models of efficacy are usually selected for their biological relevance to the disease of interest. To illustrate this point, ovalbumin-sensitized guinea pigs and Ascaris- or dust-mite-sensitized sheep have been used to evaluate the effectiveness of asthma therapies.[33–36] Guinea pigs have been used to model a wide variety of infectious diseases.[37,38] Sheep are also used to model chronic obstructive lung disease and cystic fibrosis.[39] Rat and ferret models are favored for influenza.[40–42] The capacity to knock out specific genes in mice has been universally valuable for studying genetic diseases; the ability to knock out the Cystic Fibrosis Transmembrane Receptor (CFTR) and Epithelial Sodium Channel (ENaC) receptors has allowed therapies for cystic fibrosis to be studied.[43,44]

The unique ability to study disposition of aerosolized drugs in whole organisms without risking the health of human subjects has been the overarching philosophy of this approach, as it is with all drugs and dosage forms. However, it has long been known that the unique anatomies and physiologies of different species must be considered carefully in selecting appropriate models for a particular aspect of human biology.[45–47] Frequently, the models selected represent more or less sensitive organisms than humans to bracket the worst and best potential responses to the drug. The value of having a whole organism is that the system of both biological (deposition and clearance) and pharmacological (target location, biophysics, and biochemistry) responses is intact. However, differences in species responses must be considered, such as the presence or absence of metabolizing enzymes or transporters of importance to man.[48]

Rodent species are relatively inexpensive to purchase and house and may be studied in large numbers, but administering aerosols to them requires consideration of the logistics of delivery.[49] Spray instillation, insufflation, or ventilation with an air supply carrying the aerosol can be used in anesthetized animals; but due to the size of the animals, there is a physical limit to the amount of aerosol that can be delivered without affecting their health.[49,50] Conscious rodents can be exposed to aerosols through nose-only chambers. However, the dose is limited by the concentration of the aerosol reservoir from which the animal is inhaling and by pulmonary function (tidal volume and breathing rate).[51] Dose delivered can either be calculated from known parameters for a species[52] or established by comparing pharmacokinetics of insufflated drug (known dose) at a range of doses with that achieved by inhalation from an aerosol source (unknown dose).[53]

As a general principle, using larger animals such as dogs, pigs, sheep, or nonhuman primates facilitates investigation of inhaled drugs at equivalent human doses. Drug aerosols can be given directly in the same manner as human bolus (DPIs and pMDIs) or steady-state source (nebulizers) administration, and the size of the animal and its anatomy will lead to relevant lung deposition.[31,54]

CELL CULTURE

The deficiency of animal models lies in the fact that they are not human, so there will always be a species-dependent source of error. In contrast, cell culture studies can be done with cells of human origin, and the behavior of the drug aerosol will have relevance to deposition on that particular cell in vivo. The implications for airways and alveolar epithelial cells is clear, and as a consequence, a variety of lung cells have been used, in particular small airways epithelial,[55] human bronchial epithelial,[56] Calu-3 (non-small cell lung cancer),[57] and THP-1 (acute monocytic leukemia),[58] each representing a different facet of airways exposure.[59,60] To render cells in culture physiologically relevant to human airway exposure aerosol sampling, mechanisms involving the aerodynamic and physicochemical properties of the aerosol to introduce particles to cell surfaces have been adopted.[61,62] These methods introduce species relevance to the interpretation of the data but remain limited by the infrequent use and limited scope of cocultures that mimic the entire biological response in the lungs.

Using both animal models and cell cultures brings the advantages of systems biology and species relevance to any conclusions drawn.

PHARMACOKINETICS

The discipline of pharmacokinetics applied to inhaled therapeutic agents requires special consideration.[63] Most drugs delivered to the lungs are intended for local activity, and therefore the appearance of the active moiety in the systemic circulation is an indirect measure of the dose delivered and its disposition. The location of deposition in the lungs may give rise to different local pharmacokinetics, which in turn will influence the systemic pharmacokinetics. Byron first discussed the importance of regional deposition to the disposition of immediate release inhaled drugs.[64] Gonda extended these considerations to include controlled or extended release drugs in 1988.[65]

Differences in the pharmacokinetics of drugs with various physico-chemical properties have been the subject of interest. The focus on hydrophobicity and hydrophilicity has been the most significant emphasis, which may be the result of the peculiar makeup of lung lining fluid and mucus having both a large aqueous and lipid composition.[66,67]

As interest in the possibility of in vitro to in vivo correlation has increased, the clear gap in knowledge between deposition and disposition has been identified.[68] Understanding the factors governing the disposition of drug in and from the lungs requires greater emphasis on local dissolution, clearance mechanisms, permeability of barrier fluids and cells, and relevance to target (local intracellular receptor or remote systemic pharmacological site). In this context the concept of an inhaled Biopharmaceutical Classification System, similar to that for drugs delivered to the gastrointestinal tract,[69] has been explored.[70]

PHARMACODYNAMICS

The specific pharmacodynamic outcome desired is the amelioration of a symptom or treatment of the underlying cause of disease. These targets are not necessarily accessed by the same mechanism from one drug to another. The biophysical barriers have been discussed broadly in the foregoing text. At this point, the limitations of all models in predicting therapeutic effect can be identified. Since many of the targets are intracellular and may even be associated with a specific sub-cellular organelle, there are several local barriers to delivery, including the epithelium itself if the target is in another tissue (e.g., airway muscle) and a series of membranes.

Specificity is an important component of drug delivery. For agents that act at multiple sites or have multiple functions, off-target effects are an important consideration. The use of nonspecific β-adrenergic agonists resulted in both cardiac and pulmonary effects and fatalities.[71] Subsequent identification of the specific $β_2$-adrenergic agonist isolated the desired pulmonary effects.[71] Glucocorticosteroids that deposit in the throat or are transported to the systemic circulation give rise to adverse effects. The former results in local immunosuppression that leads to Candida infection, and the latter results in cortisol suppres-sion.[72] Differences in the disposition of glucocorticosteroids have been studied with the intent of minimizing off-target effects.[73]

Pulmonary delivery of systemically acting drugs requires control of aerosol deposition and pharmacokinetics to achieve the desired disease management. Small molecular weight molecules such as ergotamine to treat migraine headaches and macromolecules such as insulin to treat diabetes[74] are subject to different transport mechanisms and rates of delivery. For solid particles where residence time in the lungs is extended, it may be important to consider dissolution rate as a poten-tial factor in the appearance of drug in plasma.[75,76]

CONCLUSION

The role of drug formulation and device combinations in assuring reproducible performance of the aerosol product is the foundation for accuracy and reproducibility of biological predictions. Placing product performance in the context of in vivo measurements and computer pre-dictions of lung deposition and clearance increases the prospect of in vitro to in vivo correlation. Advances in physiologically relevant in vitro physicochemical and biological testing, lung imaging, and pre-dictive modeling of aerosol deposition, the pharmacokinetics of drug disposition, and pharmacodynamic endpoints have significantly improved the prospects of predicting clinical outcome. However, cer-tain important elements in addressing the therapeutic objective require further elucidation, including the role of subtle differences in site of deposition on dissolution, transport (local and systemic), and the location of the nominal target (e.g., receptors, pathogens). Assuming all significant variables were known and could be controlled or accounted for, it should be possible to establish the limits to predictive modeling. The confounding of variations across all factors that are involved in drug delivery to the target site may mean that there are practical limits to the precision that can be achieved. These limits will influence the extent to which expectations of rational design can be used to streamline the product development process to accommodate the regulatory requirements and allow for rapid and less expensive treatment options for each disease.

REFERENCES

1. Byron P, et al. In vivo−in vitro correlations: Predicting pulmonary drug deposition from pharmaceutical aerosols. *J Aerosol Med Pulmonary Drug Delivery*. 2010;23(S2):S59−S69.

2. Adams W, et al. Demonstrating bioequivalence of locally acting orally inhaled drug products (OIPs): Workshop summary report. *J Aerosol Med Pulmonary Drug Delivery*. 2010;23:1−29.

3. Melani A. Inhalatory therapy training: a priority challenge for the physician. *Acta Biomed*. 2007;78:233−245.

4. Dolovich M, Dhand R. Aerosol drug delivery: developments in device design and clinical use. *Lancet*. 2011;377:1032−1045.

5. Dolovich M, et al. Consensus statement: aerosol and delivery devices. American Association for Respiratory Care. *Respir Care*. 2000;45:589−596.

6. Clark A, Hollingworth A. The relationship between powder inhaler resistance and peak inspiratory conditions in healthy volunteers—implications for in vitro testing. *J Aerosol Med*. 1993;6:99−110.

7. Olsson B, Asking L. Critical aspects of the function of inspiratory flow driven inhalers. *J Aerosol Med*. 1994;7(S1):S43−S47.

8. Dunbar C, et al. A comparison of dry powder inhaler dose delivery characteristics using a power criterion. *PDA J Pharm Sci Tech*. 2000;54:478−484.

9. Apiou-Sbirlea G, et al. Bioequivalence of inhaled drugs: fundamentals, challenges and perspectives. *Ther Deliv*. 2013;4:343−367.

10. Hickey A. Summary of common approaches to pharmaceutical aerosol administration. In: Hickey A, ed. *Pharmaceutical Inhalation Aerosol Technology*. Second ed. New York: Marcel Dekker; 2004:385−421.

11. Conway J. Lung imaging – two dimensional gamma scintigraphy, SPECT, CT and PET. *Adv Drug Deliv Rev*. 2012;64:357−368.

12. Chen D, Kinahan P. Multimodality molecular imaging of the lung. *J Magn Reson Imaging*. 2010;32:1409−1420.

13. Ishfaq M, et al. A simple radioaerosol generator and delivery system for pulmonary ventilation studies. *Eur J Nucl Med*. 1984;9:141−143.

14. Apiou-Sbirlea G, et al. Validated three dimensional CFD modeling of aerosol drug deposition in humans – influence of disease and breathing regimes. In: Dalby R, et al., eds. *Respiratory Drug Delivery*. Richmond VA: Virginia Commonwealth University; 2008:185–196.

15. Greenblatt E, et al. What causes uneven aerosol deposition in the bronchoconstricted lung? A quantitative imaging study, *J Aerosol Med Pulmonary Drug Delivery*. 2016;29:57–75.

16. Rudolf G. A mathematical model for the deposition of aerosol particles in the human respiratory tract. *J Aerosol Sci*. 1984;15:195–199.

17. Stahlhofen W, Rudolf G, James A. Intercomparison of experimental regional aerosol deposition data. *J Aerosol Sci*. 1984;2:285–316.

18. Dunbar, C., et al., *In vitro and in vivo dose delivery characteristics of large porous particles for inhalation*. Int J Pharm. 2002;245: p. 179–189.

19. Newman S, Chan H-K. In vitro/in vivo comparisons in pulmonary drug delivery. *J Aerosol Med Pulmonary Drug Delivery*. 2008;21:77–84.

20. Dunbar C, Mitchell J. Analysis of cascade impactor mass distributions. *J Aerosol Med*. 2005;18:439–451.

21. Gonda I. Study of the effects of polydispersity of aerosols on regional deposition in the respiratory tract. *J Pharm Pharmacol*. 1981;33:52P.

22. Martonen T, et al. Lung models: strengths and limitations. *Respir Care*. 2000;45:712–736.

23. Heyder J, et al. Deposition of particles in the human respiratory tract in the size range 0.005–15 μm. *J Aerosol Sci*. 1986;17:811–825.

24. Zhang Y, Chia T, Finlay W. Experimental measurement and numerical study of particle deposition in highly idealized mouth–throat models. *Aerosol Sci Technol*. 2006;40:361–372.

25. Ruzycki C, et al. Comparison of in vitro deposition of pharmaceutical aerosols in an idealized child throat with in vivo deposition in the upper respiratory tract of children. *Pharm Res*. 2014;31:1525–1535.

26. Zhou Y, Sun J, Cheng Y-S. Comparison of deposition in the USP and physical mouth-throat models with solid and liquid particles. *J Aerosol Med Pulmonary Drug Delivery*. 2011;24:277–284.

27. Olsson B, et al. Validation of a general in vitro approach for prediction of total lung deposition in healthy adults for pharmaceutical inhalation products. *J Aerosol Med Pulmonary Drug Delivery*. 2013;26:355–369.

28. Bergstrom L, Olsson B, Thorsson L. Degree of throat deposition can explain the variability in lung deposition of inhaled drugs. *J Aerosol Med*. 2006;19:473–483.

29. Mortensen N, Hickey A. Targeting inhaled therapy beyond the lungs. *Respiration*. 2014;88:353–364.

30. Beaumont C, et al. Human absorption, distribution, metabolism and excretion properties of drug molecules: a plethora of approaches. *Br J Clin Pharmacol*. 2014;78:1185–1200.

31. Pauluhn J, Mohr U. Inhalation studies in laboratory animals—current concepts and alternatives. *Toxicol Pathol*. 2000;28:734–753.

32. Wolff R, Dorato M. Toxicological testing of inhaled pharmaceutical aerosols. *Crit Rev Toxicol*. 1993;23:343–369.

33. Meurs H, et al. A guinea pig model of acute and chronic asthma using permanently instrumented and unrestrained animals. *Nat Protoc*. 2006;1:840–847.

34. Meeusen E, et al. Sheep as a model species for the study and treatment of human asthma and other respiratory diseases. *Drug Discovery Today: Dis Models*. 2009;6:101–106.

35. Bischof R, et al. Induction of allergic inflammation in the lungs of sensitized sheep after local challenge with house dust mite. *Clin Exp Allergy.* 2003;33:367–375.

36. Lowe A, et al. Adjustment of sensitisation and challenge protocols restores functional and inflammatory responses to ovalbumin in guinea-pigs. *J Pharmacol Toxicol Methods.* 2015;72:85–93.

37. Padilla-Carlin D, McMurray D, Hickey A. The guinea pig as a model of infectious diseases. *Comp Med.* 2008;58:324–340.

38. Hickey A. Guinea pig model of infectious disease – viral infections. *Curr Drug Targets.* 2011;12:1018–1023.

39. Abraham W. Modeling of asthma, COPD and cystic fibrosis in sheep. *Pulmonary Pharmacol Ther.* 2008;21:743–754.

40. Boukhvalova M, Prince G, Blanco J. The cotton rat model of respiratory viral infections pathogenesis and immunity. *Biologicals.* 2010;37:152–159.

41. Bouvier N, Lowen A. Animal models for influenza virus pathogenesis and transmission. *Viruses.* 2010;2:1530–1563.

42. Maher J, DeStefano J. The ferret: an animal models to study influenza virus. *Lab Anim.* 2004;33:50–53.

43. Zhou Z, et al. The ENaC-overexpressing mouse as a model of cystic fibrosis lung disease. *J Cyst Fibros.* 2011;10(Suppl 2):S172–S182.

44. Clarke L, et al. Defective epithelial chloride transport in a gene-targeted mouse model of cystic fibrosis. *Science.* 1992;257:1125–1128.

45. Wolff R. Experimental investigation of deposition and fate of particles: animal models and interspecies differences. In: Marijnissen J, Gradon L, eds. *Aerosol Inhalation: Research Frontiers.* Dordrecht: Springer; 1996:247–263.

46. Snipes M. Long-term retention and clearance of particles inhaled by mammalian species. *Crit Rev Toxicol.* 1989;20:175–211.

47. Snipes M. Species comparisons for pulmonary retention of inhaled particles. In: McClellan R. Henderson R. Eds. *Concepts in Inhalation Toxicology,* New York. 1989: 193–227.

48. Olsson B, et al. Pulmonary drug metabolism, clearance, and absorption. In: Smyth H, Hickey A, eds. *Controlled Pulmonary Drug Delivery.* New York: Springer; 2011:21–50.

49. Cryan S, Sividas N, Garcia-Contreras L. In vivo animal models for drug delivery across the airway mucosal barrier. *Adv Drug Deliv Rev.* 2007;59:1133–1151.

50. Durham P, et al. Disposable dosators for pulmonary insufflation of therapeutic agents to small animals. *J Vis Exp.* 2017;121(March). JoVE 55356.

51. Chaffee V. Surgery of laboratory animals. In: Melby Jr E, Altmann N, eds. *Handbook of Laboratory Animal Science.* Cleveland, OH: CRC Press; 1974:233–273.

52. Suarez S, et al. The influence of suspension nebulization or instillation on particle uptake by guinea pig alveolar macrophages. *Inhal Toxicol.* 2001;13:773–788.

53. Fiegel J, et al. Preparation and in vivo evaluation of a dry powder for inhalation of capreomycin. *Pharm Res.* 2008;25:805–811.

54. Coffman R, Hessel E. Nonhuman primate models of asthma. *J Exp Med.* 2005;201:1875–1879.

55. Benam K, et al. Small airway-on-a-chip enables analysis of human lung inflammation and drug responses in vitro. *Nat Meth.* 2015;13:151–157.

56. Forbes B, Ehrhardt C. Human respiratory epithelial cell culture for drug delivery applications. *Eur J Biopharm.* 2005;60:193–205.

57. Florea B, et al. Drug transport and metabolism characteristics of the human airway epithelial cell line Calu-3. *J Control Release*. 2003;21:131–138.

58. Carvalho CdS, Daum N, Lehr C-M. Carrier interactions with the biological barriers of the lung: advanced in vitro models and challenges for pulmonary drug delivery. *Adv Drug Deliv Rev*. 2014;75:129–140.

59. Sporty J, Horalkova L, Ehrhardt C. In vitro cell culture models for the assessment of pulmonary drug disposition. *Expert Opin Drug Metab Toxicol*. 2008;4:333–345.

60. Meindl C, et al. Permeation of therapeutic drugs in different formulations across the airway epithelium in vitro. *PLoS ONE*. 2015;10:e0135690.

61. Kazantseva M, Cooney D, Hickey A. Development of a lung model utilizing human alveolar epithelial cells for evaluating aerosol drug delivery. In: Byron P, et al., eds. *Respiratory Drug Delivery VIII*. Godalming, Surrey, UK, Raleigh, NC, USA: Davis Horwood International Publishing, Ltd; 2002:707–710.

62. Cooney D, Hickey A. Cellular response to the deposition of diesel exhaust particle aerosols onto human lung cells grown at the air–liquid interface by inertial impaction. *Toxicol In Vitro*. 2011;25:1953–1965.

63. Suarez S, Hickey A. Pharmacokinetic and pharmacodynamic aspects of inhaled therapeutic agents. In: Martonen T, ed. *Medical Applications of Computer Modeling: The Respiratory System*. Southampton, UK: WIT Press; 2001:225–304.

64. Byron P. Prediction of drug residence times in regions of the human respiratory tract following aerosol inhalation. *J Pharm Sci*. 1986;75:433–438.

65. Gonda I. Drugs administered directly into the respiratory tract: modeling of the duration of effective drug levels. *J Pharm Sci*. 1988;77:340–346.

66. Barrow R, Hills B. Properties of four lung surfactants and their mixtures under physiological conditions. *Respir Physiol*. 1983;51:79–93.

67. Frohlich F, et al. Measurements of deposition, lung surface area and lung fluid for simulation of inhaled compounds. *Front Pharmacol*. 2016;7:181.

68. Lee S, et al. In vitro considerations to support bioequivalence of locally acting drugs in dry powder inhalers for lung diseases. *AAPS J*. 2009;11:414–423.

69. Amidon G, et al. A theoretical basis for a biopharmaceutic drug classification: the correlation of in vitro drug product dissolution and in vivo bioavailability. *Pharm Res*. 1995;12:413–420.

70. Hastedt J, et al. Scope and relevance of a pulmonary biopharmaceutical classification system AAPS/FDA/USP Workshop March 16–17th, 2015 in Baltimore, MD. *AAPS Open*. 2016;2:1.

71. Hickey A. Pulmonary drug delivery: pharmaceutical chemistry and aerosol technology. In: Wang B, Hu L, Siahaan T, eds. *Drug Delivery Principles and Applications*. Second ed. New York: John Wiley and Sons, Inc.; 2016:186–206.

72. Derendorf H, et al. Relevance of pharmacokinetics and pharmacodynamics of inhaled corticosteroids to asthma. *Eur Respir J*. 2006;28:1042–1050.

73. Padden J, Skoner D, Hochhaus G. Pharmacokinetics and pharmacodynamics of inhaled glucocorticoids. *J Asthma*. 2008;45(S1):13–24.

74. Hickey A. Back to the future: Inhaled drug products. *J Pharm Sci*. 2013;102:1165–1172.

75. Gray V, et al. The inhalation ad hoc advisory panel for the USP performance tests of inhalation dosage forms. *Pharm Forum*. 2008;34:1068–1074.

76. Riley T, et al. Challenges with developing in vitro dissolution tests for orally inhaled products (OIPs). *AAPS PharmSciTech*. 2012;13:978–989.

CHAPTER 5

Regulatory Strategy

The development of an inhaled product once the drug candidate has been identified follows a series of steps, as described in earlier chapters, to prepare a formulation, combine it with a device, conduct appropriate analysis, and evaluate it in preclinical models and clinical trials.[1−4] After the initial feasibility and proof of concept studies, the steps in the process are guided by regulatory requirements for specific submissions.

Fig. 1.1 in Chapter 1 illustrates the development path and the attrition of candidate drugs that occurs at each step, initially due to scientific or technical barriers and later largely due to quality or safety questions.[5] Conducting studies according to regulatory guidance and within approved specifications constrains late stage development but also gives a pathway to approval.[2] The timelines are approximate and can be significantly abbreviated for generic drugs that are based on a product in commerce. It is evident that the time and expense involved in the development of an inhaled drug product are considerable, and in most cases the risk is much greater than that for dosage forms delivered by traditional routes of administration. Horhota and Leiner have written an outstanding account of the complex regulatory landscape that exists not only across international borders but also across agencies within countries that can make navigation to adequately address needs extremely difficult.[6]

INTERNATIONAL REGULATIONS

Each country has a regulatory authority that governs drug products in commerce within its geographical borders. In the United States, it is the Food and Drug Administration (FDA). The European Union has a transnational agreement whereby member states are subject to the directives of the European Medicines Agency. With the prospective departure of the United Kingdom from the EU, the British Medicines and Healthcare Products Regulatory Agency may play a more significant role in that country in the future. The International Council for Harmonization of Technical Requirements for Pharmaceuticals for Human Use (ICH),

Inhaled Pharmaceutical Product Development Perspectives. DOI: https://doi.org/10.1016/B978-0-12-812209-9.00005-1

founded in 1990, is increasingly important in the global harmonization of standards that ensure the quality and safety of drug products.

The use of pharmacopeias containing methods and procedures that are legally binding with respect to drug product manufacture is important in maintaining the quality of products globally. Small countries may adopt the United States and European Pharmacopeias as their primary standards documents.

In this chapter, discussion will be limited to the regulations governing drug products in the United States.

UNITED STATES REGULATIONS

The Federal Food, Drug, and Cosmetic Act of 1938 was the first to regulate the quality of drug products in the United States.[7–9] The Drug Price Competition and Patent Term Restoration Act of 1984, known as the Hatch–Waxman Act, introduced regulations to encourage generic product competition in the market.[10,11] This was intended to ensure a return on investment to innovator companies by extending patent protection while reducing cost by allowing generic competition at the end of the patent life.

In general terms, initial studies of the safety and efficacy of a product are conducted under an Investigational New Drug (IND) application.[12] The data generated under an IND can become part of a New Drug Application (NDA) or Abbreviated New Drug Application (ANDA, for generics) to the regulatory authorities.

The NDA has extensive requirements. The Code of Federal Regulations, Title 21 (CFR 21), 505(b)(1) pathway allows for the review of a drug that has not previously been approved by the FDA.[13] Drugs in existing approved products may appear in generic form through an ANDA. The CFR 21, 505(j) pathway allows for the review of a generic drug referencing a previously approved NDA.[14] The CFR 21, 505(b)(2) pathway allows for the review of a previously approved drug that as submitted differs from that of the previous approval. This pathway was described in a 1999 guidance document.[15]

Dry powder inhalers (DPIs), pressurized metered dose inhalers (pMDIs), and soft mist inhalers are products in which the drug formulation, metering system, and device are considered as an inseparable unit to be evaluated as such with respect to data for the regulatory submission. The regulatory review is done by the Center for Evaluation and Research of the FDA. Nebulizers, in contrast, are usually sold separately from the nebules containing the drug formulation and consequently are regulated primarily as medical devices. The regulatory review of nebulizers is led by the Center for Devices and Radiological Health of the FDA.

Inhaled drug products follow the general regulatory pathways described above, but there are unique considerations when contemplating delivery systems that are equivalent to those already in commerce. The complexity of manufacturing and ensuring the quality and performance characteristics of a drug-formulation-device combination with respect to an existing product makes the 505(j) pathway much more difficult than the 505(b)(2) approach. Consequently, generic drug manufacturers may pursue the latter. However, there is a marketing disadvantage to this approach in that a product approved by the 505(j) process is automatically substitutable for the innovator product in pharmacies. This is not automatically the case for those products approved by the 505(b)(2) pathway. Consequently, the initial market penetration by the 505(b)(2) product is likely to be smaller than that of the 505(j).

Where the intention is to obtain regulatory approval for a new nebulizer, the usual approach is the 510(k). This allows demonstration that the new device is substantially equivalent to a legally marketed device under 21CFR.

Regulated studies are supported by cGXP procedures (current good laboratory,[16–18] manufacturing,[19] and clinical[20,21] practices). Preclinical studies are reviewed by institutional animal care and use committees (IACUCs) governed by the Association for Assessment and Accreditation of Laboratory Animal Care and the National Institutes of Health Office of Laboratory Animal Welfare accredited methods.[22,23] Clinical studies are reviewed by institutional review boards (IRBs) at the study site.[24] Fig. 5.1 illustrates the numbers of patients that may be involved at different phases of the clinical trials. At the investigational stage, a small number of healthy volunteers and patients may be required. The number increases as the product moves through Phases II and III to potentially very large numbers

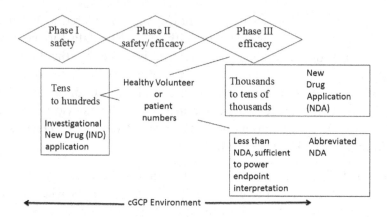

Figure 5.1 Clinical trials design and patient numbers.

for new drugs and usually a more modest number for those going through an abbreviated application, in which a comparator product is available.

GENERAL PRINCIPLES

As understanding of drug products and the sophistication of technology has increased, parallel developments in regulatory strategy have occurred. Since the goals of regulation are primarily to establish the quality and safety of drug products, ensuring the accuracy and reproducibility of dose delivery is the basis for all methods of evaluation. Historically, discrete controls were placed on processes and procedures, starting with the product manufacture and characterization and extending to the performance of clinical trials. Statistical methods were used to set specifications and monitor controls with the ability to dispose of batches of materials from unit processes or based on clinical studies that were defined as not meeting the desired specification.

In recent years, progress in process analytical technology and on-line and in-line monitoring has given immediate control over manufacturing processes, limiting the extent of quality testing required and increasing the prospects of the product meeting specification.[25] The two most important guiding principles that have been promulgated by the FDA are quality by design (QbD) manufacturing and weight of evidence characterization.

Quality by Design

QbD methods are underpinned by statistically designed experiments and statistical process control. The use of multivariate statistics for pharmaceutical process development is relatively recent despite the long history of use in general engineering.[26] Many pharmaceutical manufacturing processes are characterized by a complex combination of input variables each of which alone or in combination can result in variances in the quality or performance of the dosage form and drug product. To fully evaluate the process space, that combination of variables used for product manufacture, it is necessary to establish a region in which output variables have the least sensitivity to input variables, thereby rendering them unlikely to be affected by modest changes in the conditions that might occur over time. Moreover, it is important to assess the role of environmental factors, as there may be fluctuations in temperature, humidity, sound, or other mechanical sources of energy input or dissipation that have an impact on the process. The FDA produced a guidance document on the application of QbD principles to the inhaled drug product development to promote the use of rational methods and control strategies that will ensure the final product performance.[27]

Weight of Evidence

The weight-of-evidence approach has been applied in several FDA settings for drug product evaluation, most notably in toxicology.[28] Not all observations in toxicological studies are associated with toxicity, but as observations accumulate, particularly those of overtly adverse events, the balance of opinion on the safety of the product may tip against further development.

It is important to note that inhalation toxicology requires testing of the drug product administered by the pulmonary route with the specific intent of identifying any local effects. In addition, systemic effects will be observed, but these may also be derived from independent studies of the drug administered by other routes.

In the case of inhaled insulin, several observations were made that initially raised concerns about the approvability of the product, including fibrosis in rats, an increase in insulin antibodies, and a drop in PO_2.[29–31] However, all of these observations were explained, and ultimately none was sufficient to impede the approval of Exubera in 2007.[32,33]

Chemistry Manufacturing and Controls

The starting materials—drug, excipients, and device components— are accompanied by drug manufacturing files and certificates of analysis ensuring the quality of chemicals by the manufacturer or supplier.

The chemistry manufacturing and controls section of submissions to the FDA was described in a draft guidance document almost 20 years ago.[34] While it has not been published as a guidance document, it functions in that role until it is either moved to final status or revised and updated. It is complemented by the United States Pharmacopeia (USP) General Chapters on quality <5> and performance <601> tests and dosage form description <1151>.[35]

General chapter <5> describes product quality attributes that in most cases are similar to those of other dosage forms, including identification assay, degradation products to assure identification, and the strength, quality, purity, and potency of the drug product. Other important measures are microbial content, foreign materials, leachables, and water content. Spray pattern is a product-specific test for products that deliver drug independent of inspiratory flow, particularly pMDIs.

General chapter <601> focuses on delivered dose uniformity (DDU) and aerodynamic particle size distribution (APSD) as key product performance measures, which are discussed in Chapters 3 and 4. It should be noted that tests are described for all currently marketed nasal and inhalation products. Dosage-form-specific instruments and apparatus assemblies are required to address the mechanisms of aerosol delivery.

Nebulizers are not sold with the formulation as a drug product. Consequently, the USP does not have requirements for testing and specifications. However, general chapter <1601> is an informational document that describes tests that are suitable for the performance of these devices in combination with specific drug formulations and complements general chapter <601> in that it describes DDU and APSD measurement. Since nebulizer aerosol delivery is mostly a steady state phenomenon, where the dose is delivered in a period of roughly 10 to 30 minutes, depending on the system, the delivery rate is also an important consideration.

Product-Specific Guidance Documents

Fig. 5.2 illustrates the range of activities involved in the development of products up to full-scale manufacturing. The intent of the figure is simply to show the sorts of things that might be involved and that contribute to the complexity of product development.

A trend for the publication of guidance on specific products to encourage generic product submissions was initiated in 2013 with two new documents by the FDA. These guidance documents described an albuterol sulfate product and a fluticasone dipropionate and salmeterol xinafoate combination product.[36,37] These were followed by 14 product-specific guidances in 2015 and 2016.[36,38-51] Table 5.1 identifies each.

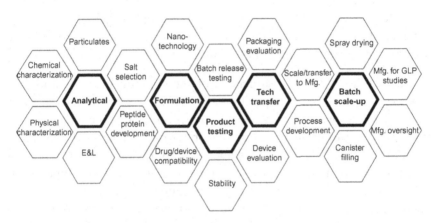

Figure 5.2 Drug product technical development activities.

Table 5.1 US Food and Drug Administration Drug-Specific Draft Guidances		
Metered Dose Inhalers	**Dry Powder Inhalers**	**Nebulizers**
2016 albuterol sulfate (revised from 2013)[36] beclomethasone dipropionate[41] budesonide and formoterol fumarate[42] ciclesonide[43] formoterol fumarate and mometasone furoate[44] mometasone furoate[45]	2016 budesonide[46] fluticasone furoate[51] fluticasone furoate and vilanterol trifenetate[47] indacaterol maleate[48] umeclidinium bromide[49]	2012 budesonide[52]
2015 ipratropium bromide[50] levalbuterol tartrate[38]	2015 aclidinium bromide[39] formoterol fumarate[40]	
	2013 fluticasone dipropionate and salmeterol xinafoate[37]	

As these guidance documents are approved, the questions of bio-equivalence and pharmaceutical equivalence addressed in Chapter 4 become increasingly important. The concept of essential similarity is central to the approach adopted by regulatory authorities for the development of generic products. The FDA uses terminology in defining bioequivalence as follows: Q1 for qualitatively the same, Q2 for quantitatively the same, and Q3 for similar physicochemical attributes of the specific dosage form. To accommodate these requirements, the test and reference products contain the same active and inactive ingredients (Q1), the same amounts of active and inactive ingredients (Q2), and have the same physicochemical properties. These abbreviations are not to be confused with the ICH quality guidelines Q1–Q12 that cover the broad range of considerations necessary to define the quality of a product throughout its lifespan. The ICH quality guidelines are not pertinent to the present discussion but can readily be accessed at www.ich.org/products/guidelines/quality.html.

The similarity (ideally sameness) of the components (Q1) and their assembly (e.g., blend composition, suspension or solution concentration, Q2) is a clear prerequisite for the product to be employed. It is then subjected to in vitro and in vivo (pharmacokinetic and pharmacodynamics) studies, which are somewhat abbreviated, to demonstrate similarity to the innovator performance. Interestingly, and uniquely to the US market, the concept of pharmaceutical equivalence is layered onto the measures that are designed to assure bioequivalence. Pharmaceutical equivalence demands that the generic product appear and operate similarly to the innovator product. This is a modest barrier for pMDIs where the components are commodities and all devices appear very similar. However, for DPIs where the innovator has patents covering the formulation and device design, it may be difficult to match the device exactly, and, since some factors in the design have both form and function elements, demonstrating bioequivalence may also be challenging.

Public Collaboration and Private Consortium

To promote excellence in the scientific and technical foundation for regulation, a public collaboration and a private consortium have published white papers on specific issues of importance to orally inhaled drug products. The Product Quality Research Institute, composed of representatives from regulatory agencies, industry, professional societies and academia, has expended considerable effort in illuminating the methods associated with aerosol testing.[53] The International Pharmaceutical Aerosol Consortium on Regulation and Science has an outstanding record of addressing important manufacturing and analytical considerations with respect to monitoring and control of inhaled drug products to ensure quality and safety.[54] The literature produced by these groups is extensive and topic specific. While it is beyond the scope of this overview to cite the literature, it should be noted when considering regulatory strategy that publications from these groups are very informative and enormously helpful.

CONCLUSION

A large number of candidate molecules are identified in the pharmaceutical discovery phase. As they proceed through early development, many are found to have undesirable physicochemical properties or dosing limitations that preclude further development. Assuming that the compound can be formulated into a product in a manner that is suitable for scale-up through process engineering, it then enters preclinical and clinical testing (animal studies and Phase I human studies), where the initial focus is on safety. For inhaled products, safety testing must consider the pulmonary route of administration in addition to the systemic effects. Products that show no significant adverse effects proceed to efficacy testing (Phases II and III human studies[7]), the last barrier to showing efficacy for either a new indication or an old indication. In the latter case, the additional comparison of equal or superior efficacy to an existing product may be a consideration.

Throughout this process, ethical and regulatory bodies monitor the design and implementation of experiments and manufacturing procedures. IACUCs and IRBs monitor the conduct of animal and human studies, respectively. Regulators review the IND as the compound is entering clinical trials, and subsequently the NDA or ANDA. The facilities in which a drug is manufactured are subject to Current Good Manufacturing Practice regulations, animal studies under Current Good Laboratory Practice regulations, and clinical trials under Current Good Clinical Practice regulations. The processes and procedures are subject to review at any time by the regulatory body, and onsite inspections occur routinely to ensure adherence.

Guidance documents and pharmacopeial standards have been established to support the pharmaceutical manufacturer in product development activities. General guidances and chapters are available, and recently product-specific guidances have been published. The latter are intended to facilitate generic product submissions. This chapter summarizes the items of interest to regulators; more details on methods can be found in earlier chapters.

Collaboration by a wide range of stakeholders has resulted in a background of excellent white papers on issues of current regulatory importance, from which researchers in the field can supplement the formal documents issued by regulators to assist in assembling a regulatory strategy for their particular inhaled drug product.

REFERENCES

1. Hickey A, Misra A, Fourie P. Dry powder antibiotic aerosol product development: inhaled therapy for tuberculosis. *J Pharm Sci.* 2013;102:3900–3907.

2. Singh G, Poochikian G. Development and approval of inhaled respiratory drugs: a US regulatory perspective. In: Smyth H, Hickey A, eds. *Controlled Pulmonary Drug Delivery.* New York: Springer; 2011:489–527.

3. Petrova E. Innovation in the pharmaceutical industry: the process of drug discovery and development. In: Ding M. Eliashberg J. Stremersch S, Eds. *Innovation and Marketing in the Pharmaceutical Industry*, Springer: New York. 19–81.

4. McElroy M, et al. Inhaled biopharmaceutical drug development: nonclinical considerations and case studies. *Inhal Toxicol.* 2013;25:219–232.

5. http://www.PhRMA.org.

6. Horhota S, Leiner S. Developing performance specifications for pulmonary products. In: Smyth H, Hickey A, eds. *Controlled Pulmonary Drug Delivery.* New York: Springer; 2011:529–541.

7. The Food, *Drug and Cosmetic Act, Pub. L. No. 75-717, ch. 675, 52 Stat. 1040 (June 25, 1938) (codified as amended at 21 U.S.C. §§ 301–399 (2002). at 21 U.S.C. § 355 (2006).*

8. US Food and Drug Administration, *The 1938 Food, Drug and Cosmetic Act—FDA.* <https://www.fda.gov/aboutfda/whatwedo/history/productregulation/ucm132818.htm>; 2009.

9. US Food and Drug Administration, *1938 Food, Drug and Cosmetic Act—FDA.* <https://www.fda.gov/aboutfda/whatwedo/history/origin/ucm054826.htm>; 2012.

10. The Drug *Price Competition and Patent Term Restoration (Hatch–Waxman) Act, Pub. L. No. 98-417, 98 Stat. 1585 (1984) (codified as amended at 21 U.S.C. §355 and 35 U.S.C. §156, 271 and 282.*

11. Rumore M. *The Hatch–Waxman Act—25 years later: keep the pharmaceutical scales balanced. Pharmacy Times,* Generic Supplements: August. 2009;4–7, 20.

12. Holbein M. Understanding FDA regulatory requirements for investigational new drug applications for sponsor-investigators. *J Investig Med.* 2009;57:688–694.

13. Lee K, Bacchetti P, Sim I. Publication of clinical trials supporting successful new drug applications: a literature analysis. *PLoS Med.* 2008;5:e191.

14. Lee S, et al. Regulatory considerations for approval of generic inhalation drug products in the US, EU, Brazil, China and India. *AAPS J.* 2015;17:1285–1304.

15. US Food and Drug *Administration, 505(b)(2) Approval Pathway.* <https://www.fda.gov/downloads/Drugs/Guidances/ucm079345.pdf>; 1999.

16. Vijayaraghavan R, et al. GLP (Good Laboratory Practice) guidelines in academic and clinical research: ensuring protection and safety. *J Pharm Res Clin Pract.* 2014;4:89–104.

17. Wedlich R, et al. Good laboratory practice. Part 1. An introduction. *J Chem Educ.* 2013;90:854–857.

18. Pesez M. Good laboratory practice in pharmaceutical quality control. *J Pharm Biomed Anal.* 1983;1:385–391.

19. Tobin E. *An Introduction to cGxP and Validation for Engineers.* Second ed Waterford, Ireland: SoloValidation Resources Limited; 2015.

20. Vijayananthan A, Nawawi O. The importance of Good Clinical Practice guidelines and its role in clinical trials. *Biomed Imaging Interv J.* 2008;4:e5.

21. Woolf S, et al. Potential benefits, limitations and harms of clinical guidelines. *BMJ.* 1999;318:527–530.

22. *AAALAC Rules of Accreditation.* <https://www.aaalac.org/accreditation/rules.cfm>.

23. *Office of Laboratory Animal Welfare, National Institutes of Health.* <https://grants.nih.gov/grants/olaw/olaw.htm>.

24. Jacobs M. Institutional review boards and independent ethics committees. In: McGraw M, et al., eds. *Principles of Good Clinical Practice.* Gurnee, IL: Pharmaceutical Press; 2010:121–147.

25. Hickey A, Ganderton D. Process analytical technology. In: Hickey A, Ganderton D, eds. *Pharmaceutical Process Engineering.* Second ed. New York: Informa Healthcare; 2010:202–204.

26. Hickey A, Ganderton D. Statistical experimental design. In: Hickey A, Ganderton D, eds. *Pharmaceutical Process Engineering.* Second ed. New York: Inform Healthcare; 2010:197–201.

27. Hickey A, Ganderton D. Quality by design. In: Hickey A, Ganderton D, eds. *Pharmaceutical Process Engineering.* Second ed. New York: Informa Healthcare; 2010:193–196.

28. Cormier J. Advancing FDA's regulatory science through weight of evidence evaluations. *J Contemp Health Law Policy.* 2011;28. Article 2.

29. Selam J-L. Inhaled insulin: Promises and concerns. *J Diabetes Sci Technol.* 2008;2:311–315.

30. Stoever J, Palmer J. Inhaled insulin and insulin antibodies: a new twos to an old debate. *Diabetes Technol Ther.* 2002;4:157–161.

31. Valente A, et al. Recent advances in the development of an inhaled insulin product. *BioDrugs.* 2003;17:9–17.

32. Quattrin T, et al. Efficacy and safety of inhaled insulin (Exuber) compared with subcutaneous insulin therapy in patients with type I diabetes. *Diabetes.* 2004;27:2622–2627.

33. Barnett A. Exubera inhaled insulin: a review. *Int J Clin Pract.* 2004;58:394–401.

34. US Food and Drug *Administration, Metered Dose and Dry Powder Inhaler Guidance.* <http://www.fda.gov/downloads/Drugs/GuidanceComplianceRegulatoryInformation/Guidances/UCM70573.pdf>; 1998.

35. *United States Pharmacopeia.* USP 40, NF 45.

36. US Food and Drug *Administration, Albuterol Sulfate Guidance.* <http://www.fda.gov/downloads/Drugs/GuidanceComplianceRegulatoryInformation/Guidances/UCM346985. pdf>; 2016.

37. US Food and Drug *Administration, Fluticasone Dipropionate and Salmeterol Xinafoate Guidance.* <http://www.fda.gov/downloads/Drugs/GuidanceComplianceRegulatoryInformation/Guidances/ UCM367643.pdf>; 2013.

38. US Food and Drug *Administration, Levalbuterol Tartrate Guidance.* <http://www.fda.gov/ downloads/Drugs/GuidanceComplianceRegulatoryInformation/Guidances/UCM452780.pdf>; 2015.

39. US Food and Drug *Administration, Aclidinium Bromide Guidance.* <http://www.fda.gov/down-loads/Drugs/GuidanceComplianceRegulatoryInformation/Guidances/UCM460918.pdf>; 2015.

40. US Food and Drug *Administration, Formoterol Fumarate.* <http://www.fda.gov/downloads/ Drugs/GuidanceComplianceRegulatoryInformation/Guidances/UCM461064.pdf>; 2015.

41. US Food and Drug *Administration, Belomethasone Dipropionate Guidance.* <http://www.fda.gov/ downloads/Drugs/GuidanceComplianceRegulatoryInformation/Guidances/UCM481768.pdf>; 2016.

42. US Food and Drug *Administration, Budesonide and Formoterol Fumarate Guidance.* <http:// www.fda.gov/downloads/Drugs/GuidanceComplianceRegulatoryInformation/Guidances/ UCM452690.pdf>; 2016.

43. US Food and Drug *Administration, Ciclesinide Guidance.* <http://www.fda.gov/downloads/ Drugs/GuidanceComplianceRegulatoryInformation/Guidances/UCM481787.pdf>; 2016.

44. US Food and Drug *Administration, Formoterol Fumarate and Mometasone Furoate.* <http:// www.fda.gov/downloads/Drugs/GuidanceComplianceRegulatoryInformation/Guidances/ UCM481824.pdf>; 2016.

45. US Food and Drug *Administration, Mometasone Furoate Guidance.* <http://www.fda.gov/down-loads/Drugs/GuidanceComplianceRegulatoryInformation/Guidances/UCM495387.pdf>; 2016.

46. US Food and Drug *Administration, Budesonide (DPI) Guidance.* <http://www.fda.gov/down-loads/Drugs/GuidanceComplianceRegulatoryInformation/Guidances/UCM533023.pdf>; 2016.

47. US Food and Drug *Administration, Fluticasone Furoate and Vilanterol Trifenetate Guidance.* <http://www.fda.gov/downloads/Drugs/GuidanceComplianceRegulatoryInformation/Guidances/ UCM495023.pdf>; 2016.

48. US Food and Drug *Administration, Indacaterol Maleate Guidance.* <http://www.fda.gov/ downloads/Drugs/GuidanceComplianceRegulatoryInformation/Guidances/UCM495054.pdf>; 2016.

49. US Food and Drug *Administration, Umeclidinium Bromide Guidance.* <http://www.fda.gov/ downloads/Drugs/GuidanceComplianceRegulatoryInformation/Guidances/UCM520285>; 2016.

50. US Food and Drug *Administration, Ipratropium Bromide Guidance.* <http://www.fda.gov/ downloads/Drugs/GuidanceComplianceRegulatoryInformation/Guidances/UCM436831.pdf>; 2015.

51. US Food and Drug *Administration, Fluticasone Furoate (DPI) Guidance.* <http://www.fda.gov/ downloads/Drugs/GuidanceComplianceRegulatoryInformation/Guidances/UCM495024.pdf>; 2016.

52. US Food and Drug *Administration, Budesonide (nebulized) Guidance.* <http://www.fda.gov/ downloads/Drugs/GuidanceComplianceRegulatoryInformation/Guidances/UCM319977.pdf>; 2012.

53. http://www.pqri.org.

54. http://www.ipacrs.org.

CHAPTER 6

General Conclusion

The pharmaceutical development process concludes with safety and efficacy in the target disease patient population. Generalizing about this topic is impossible as each disease presents a different set of desired outcomes and measures of effectiveness and unique potential for toxicity. Some discussion of this topic is essential as a conclusion to this volume.

DISEASES

The purpose of inhaled pharmaceutical product development is to commercialize therapeutic options in order to alleviate the symptoms or treat the underlying cause of disease. In this context, it is important to reflect on the therapies which these products deliver and the diseases they will be used to treat.

All development activities end with an efficacy or pharmacodynamic study in which a relevant clinical endpoint is measured to assess the effectiveness of the drug. In a general text of this nature, the wide range of diseases and potential clinical endpoints does not lend itself to generalities. Phase II and III clinical studies are designed uniquely to address the desired treatment outcome and specific disease.

Fig. 6.1 illustrates several of the diseases currently being treated and, more importantly, the way in which the targets are differentially distributed in the lungs according to the pathology of disease. Following delivery of the drug to the lungs, the site of deposition becomes an important driver for its disposition and action.[1,2]

The airways are the major site of action for many drugs involved in the treatment of cystic fibrosis (CF).[2] The first barrier a drug experiences on deposition is the mucus, and this is the site of action for mucolyic agents (salt, acetyl cysteine, DNAse) that are used to alleviate congestion.[3] In CF, the epithelial cell surface is the site of the genetic defect, and the transmembrane receptor (CFTR) is responsible

Inhaled Pharmaceutical Product Development Perspectives. DOI: https://doi.org/10.1016/B978-0-12-812209-9.00006-3

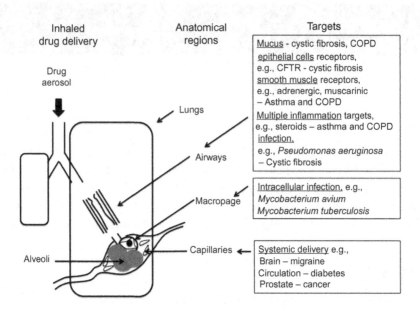

Figure 6.1 Anatomical regions in which inhaled drugs deposit and the target site, receptor, pathogen, or disease.

for chloride ion transport that results in an insufficiency of water in the airways. This is complemented by the epithelial sodium channel (ENaC). Both the CFTR and the ENaC are targets for drugs. The airways epithelium has also been a major target for gene therapy to correct expression of the CFTR.

The airways are a major target site in asthma since they are the location of the smooth muscle involved in bronchoconstriction, a symptom of the disease. Adrenergic and muscarinic agonists are delivered to this site to induce bronchodilatation through the sympathetic and parasympathetic autonomic nervous systems.[4] The underlying cause of disease, inflammation, is thought to be present throughout the lungs. Therefore, the administration of glucocorticosteroids is intended to treat both airways and periphery.[5]

Chronic obstructive pulmonary disease (COPD) is treated with many of the same drugs used to treat asthma, the most prominent of which are Advair/Seretide (fluticasone dipropionate and salmeterol xinafoate) and Spiriva (tiotropium).[6] The inflammatory component of COPD may be accompanied by infection. Infection is also an important component of the pathology of CF, where treatment of *Pseudomonas aeruginosa* infection is an important element of the remedy.[3]

Treatment of organisms that are known to grow in macrophages may require a specialized strategy. Notably, *Mycobacterium avium* and *Mycobacterium tuberculosis* have been treated with high drug dose aerosols.[7,8]

A variety of systemic diseases have been treated with inhaled aerosols, including migraine (ergotamine and hydroxyergotamine), diabetes (insulin), and prostate cancer (leuprolide acetate). Unfortunately, many of these therapies either did not reach the market or had a short period in commerce.[6]

It can be concluded from the observation of the nature of disease that the targets vary between diseases. Consequently, it is beyond the scope of this monograph to consider specific pharmacodynamics endpoints or clinical outcomes as they relate to the product in development.

DOSAGE FORMS

The major dosage forms—pressurized metered dose inhalers, dry powder inhalers, nebulizers, and soft mist inhalers—were discussed in previous chapters, and it is evident that these will remain the primary forms for the conceivable future. However, researchers in academia and industry are exploring novel delivery systems. The most interesting appear to be devices that use a new route of administration, transnasal. These systems generate aerosols that are suitable for delivery through nasal cannulae under passive breathing conditions of both liquid and powder aerosols for action in the lungs.[9,10]

While new devices are being conceived, new methods of rapid prototyping are also being developed, notably 3D printing, that will allow devices to be constructed and initial testing conducted to iterate quickly to a final design to be made by stereo lithography or traditional plastics molding.[11] 3D printing technology might advance to the point where it has production scale capability and prototypes can be linked directly to production scale devices by a single design and manufacturing process.[12]

QUALITY AND PERFORMANCE TESTS

The quality and performance tests used to characterize pharmaceutical aerosol drug products fall generally into physicochemical characterization and aerosol performance methods. Because the scientific principles on which these methods are based are not subject to change, very little technical development can be expected.

As the desire for physiologically relevant methods increases, there is an increasing need to define the point at which quality tests required to satisfy regulatory requirements to characterize the product deviate from performance tests used in development in attempts to predict performance in humans. It is not clear at this point that all approaches adopted to mimic clinical performance will translate into tools for quality assessment.

Model faces to which facemasks can be applied to test nebulizer performance[13] and physiological inlets to impactors, as described earlier, are the best examples of advancements that improve determination of delivered dose uniformity and aerodynamic particle size distribution.[14] While a case could be made for expanding consideration of aerosol delivery rate, which is currently assessed for nebulizer output, to all other inhalers, it is not yet clear that a universally applicable method is available, or that this additional information would be valuable once specifications on product performance have been set. Should such methods become available, this topic could be revisited.

PRECLINICAL AND CLINICAL CONSIDERATIONS

The ethical use of animals in efficacy and safety testing has resulted in the three R's (refine, reduce, replace) approach.[15] Possibly the single most significant change that could occur in preclinical testing in the future is movement from in vivo studies to a combination of in vitro human cell culture (2D and 3D structures) and in silico (computer models) that might in time replace animal models.[16] It is much too early to say with confidence that this will occur, but major efforts are underway to achieve this objective.

It is conceivable that evolution of preclinical testing might link directly to an in vitro, in vivo correlation approach to predicting disposition of drugs and, where clear endpoints can be identified, efficacy. The latter would require much greater understanding of the link between aerosol performance, deposition, clearance, and pharmacokinetics with efficacy.[17] As was discussed in Chapter 4, the links between in vitro aerosol testing through to pharmacokinetics are becoming increasingly clear.[18,19] Rapid developments in the field of molecular and cellular biology are underpinning identification of specific biomarkers for disease. As this field develops, it may be possible to make quantitative measures that correlate significantly with therapeutic outcome and limit the dependence on insensitive physiological outcomes.

REGULATORY STRATEGY

The regulatory approach evolves as a function of new scientific and technical developments and the unmet medical need. The examples given in Chapter 5 focus on the US Food and Drug Administration (FDA) approach to the broad topic of quality and performance tests for orally inhaled drug products, which, while not identical to those of other countries, are representative of the sorts of tests that all regulatory agencies promulgate. There has been an enormous effort to establish guidances to support generic product development; all of the recent documents were described earlier.

The observation that chlorofluorocarbon propellants were involved in ozone depletion and consequent negative health effect led to the Montreal Protocol, an international ban on their use.[18,20] As alternative hydrofluoroalkane propellants were introduced, a regulatory position on the elimination and replacement of CFCs was required. The FDA guidance on this is an example of the way in which regulators can respond to new scientific and technical developments.[21]

The recent upturn in antibiotic resistant organisms has raised a need for FDA guidance on the route to approval of new drug candidates, reflecting the demonstrable urgency. In response to this serious threat to public and global health, a guidance document has appeared that should facilitate the rapid approval of new drug candidates.[22]

NEW DISEASE TARGETS

An enormous body of knowledge, scientific and technical skill, grasp of the biology to support preclinical and clinical testing, understanding of disease pathogenesis, and project management processes for regulatory submission are now available for a wide range of diseases. As new techniques are developed, new options may emerge.

Immunology is a rapidly evolving discipline and opens up new therapeutic targets in the lungs that may help in the treatment of cancer, allergic, autoimmune, and infectious diseases. The use of host-directed therapy or immunotherapy is rapidly gaining acceptance, and it is conceivable that aerosols may help with localized disease or lung manifestations of systemic diseases.

Aerosolized gene therapy to correct the underlying genetic cause of CFTR under-expression showed early promise[23] but has languished for many years without demonstrating compelling clinical success. The recent revolution in genetic correction brought about by the gene editing technology of clustered regularly interspaced short palindromic repeats (CRISPR)[24] may bring aerosol technologies originally developed for gene therapy into development.

As a product of these factors and an indicator of their potential, new therapies are being developed for pulmonary arterial hypertension, idiopathic pulmonary fibrosis, and acute lung injury.[25-27]

CONCLUSION

A perspective on the range of activities involved in inhaled pharmaceutical product development has been presented. This small volume serves as an introduction for those new to the field while illuminating some of the most significant considerations that overlie the detailed critical path guiding product development activity for those more familiar with the subject. It is important not to lose sight of these high-level considerations when involved in the daily tasks that accompany each phase. Crucial decision points often occur at the transition between the areas described in each of the chapters. Specific diseases require unique considerations. Linking general development activities to the outcomes of safety and efficacy for a particular disease would require a far more comprehensive document and would more likely be the topic of a development plan than a short monograph.

In these early decades of the new millennium, it is clear that this century has the potential to exceed the achievements of the preceding one in promoting inhaled drug therapy. However, if we examine the technology and methods currently employed, we see that they owe more to the last than the present century. Nevertheless, there is sufficient innovation across the spectrum of development activities for optimism about the future of new technologies and the potential to address the medical needs of diseases as yet untreated with aerosols.

REFERENCES

1. Lipworth B. Targets for inhaled treatment. *Respir Med.* 2000;94(SD):S13–S16.

2. Groneberg D, et al. Fundamentals of pulmonary drug delivery. *Respir Med.* 2003;97:382–387.

3. Contreras LG, Hickey A. Aerosol treatment of cystic fibrosis. *Crit Rev Ther Drug Carrier Syst.* 2003;20:317–356.

4. Hickey A. Pulmonary drug delivery: pharmaceutical chemistry and aerosol technology. In: Wang B. Hu L. Siahaan T, Eds. *Drug Delivery Principles and Applications*; Second ed. New York, John Wiley and Sons, Inc., 2016: 186–206.

5. Townley R, Suliaman F. The mechanism of corticosteroids in treating asthma. *Ann Allergy.* 1989;58:1–6.

6. Hickey A. Back to the future: inhaled drug products. *J Pharm Sci.* 2013;102:1165–1172.

7. Olivier K, et al. Inhaled amikacin for treatment of refractory pulmonary nontuberculous mycobacterial disease. *Ann Am Thorac Soc.* 2014;11:30–35.

8. Hickey A, et al. Inhaled drug treatment for tuberculosis: past progress and future prospects. *J Controlled Release.* 2016;240:127–134.

9. Zeman K, et al. A trans-nasal aerosol delivery device for efficient pulmonary deposition. *J Aerosol Med Pulmonary Drug Delivery.* 2017;30:223–229.

10. Longest P, et al. Efficient nose-to-lung (N2L) aerosol delivery with a dry powder inhaler. *J Aerosol Med Pulmonary Drug Delivery.* 2015;28:189–201.

11. Shamah D. Breakthrough cannabis inhaler weds medical tech, 3D printing. *Start-up Israel.* 2017; (August 5th).

12. Tumbleston J, et al. Continuous liquid interface production of 3D objects. *Science.* 2015;347:1349–1352.

13. Mitchell J, Suggett J, Nagel M. Clinically relevant in vitro testing of orally inhaled products—bridging the gap between the lab and the patient. *AAPS PharmSciTech.* 2016;17:787–804.

14. Ung K, et al. In vitro assessment of dose delivery performance of dry powder inhalers. *Aerosol Sci Technol.* 2014;48:1099–1110.

15. Flecknell P. Replacement, reduction and refinement. *ALTEX.* 2002;19:73–78.

16. Ehrhard C, Kim K-J. *Drug Absorption Studies: In Situ, In Vitro and in Silico Models.* New York: Springer; 2008.

17. Apiou-Sbirlea G, et al. Bioequivalence of inhaled drugs: fundamentals, challenges and perspectives. *Therap Deliv.* 2013;4:343–367.

18. Olsson B, et al. Validation of a general in vitro approach for prediction of total lung deposition in healthy adults for pharmaceutical inhalation products. *J Aerosol Med Pulmonary Drug Delivery.* 2013;26:355–369.

19. Byron P, et al. In vivo–in vitro correlations: predicting pulmonary drug deposition from pharmaceutical aerosols. *J Aerosol Med Pulmonary Drug Delivery.* 2010;23(S2):S59–S69.

20. Molina M, Rowland F. Stratospheric sink for chlorofluoromethanes: chlorine atom-catalysed destruction of ozone. *Nat Rev Genet.* 1974;249:810–812.

21. Stein S, Thiel C. The history of therapeutic aerosols: a chronological review. *J Aerosol Med Pulmonary Drug Delivery*. 2017;30:20–41.

22. US Food and Drug *Administration, Guidance for Industry, Antibacterial Therapies for Patients with Unmet Medical Need for the Treatment of Serious Bacterial Diseases*. <https://www.fda.gov/downloads/Drugs/GuidanceComplianceRegulatoryInformation/Guidances/UCM359184.pdf>; 2017.

23. Stribling R, et al. Aerosol gene delivery in vivo. *PNAS*. 1992;89:11277–11281.

24. Zhang F, Wen Y, Guo X. CRISPR/Cas9 for genome editing: progress, implications and challenges. *Hum Mol Genet*. 2014;23:R40–R46.

25. Hill N, Preston I, Roberts K. Inhaled therapies for pulmonary hypertension. *Respir Care*. 2015;60:802–805.

26. Ivanova V, et al. Inhalation treatment of pulmonary fibrosis by liposomal prostaglandin E2. *Eur J Pharm Biopharm*. 2013;84:335–344.

27. Taylor R, Zimmerman J, Dellinger R. Low-dose inhaled nitric oxide in patients with acute lung injury. *JAMA*. 2004;291:1603–1609.

INDEX

Note: Page numbers followed by "*f*" and "*t*" refer to figures and tables, respectively.

Printed in the United States
By Bookmasters